ADVANCED QUANTUM THERMODYNAMICS (IS A SUBJECT I KNOW VERY LITTLE ABOUT)

I0464238

ADVANCED QUANTUM THERMODYNAMICS (IS A SUBJECT I KNOW VERY LITTLE ABOUT)

Selected Writings
by

DAVID NG

Materials published
from 2004 to 2014

SCIENCE CREATIVE
QUARTERLY PUBLICATIONS

Copyright 2015 David Ng

All rights reserved under International and Pan-American Copyright Conventions. Published via lulu.com

See also:
scq.ubc.ca
popperfont.net
thisishowitalkscience.tumblr.com

ISBN: 978-1-312-89413-6

CONTENTS

[a] Originally published in *McSweeneys.net*
[b] Originally published in the *Science Creative Quarterly*
[c] Originally published in *Yankee Pot Roast*
[d] Originally published in *Monkey Bicycle*
[e] Originally published in *The Believer*
[f] Originally published in *The Walrus*
[g] Originally published in *BoingBoing*
[h] Originally published in *Inkling Magazine*
[i] Originally published in *Seed Magazine*
[j] Originally published in *Rice Paper*
[k] Originally published in *Maisonneuve*

WHY I DO SCIENCE

When I look out my office window, I see two sets of nucleotide bases – guanine and cytosine. I don't mention this as an admission of psychotic delirium. The building where I work just happens to have a DNA molecule emblazoned on its windows. Admittedly, it's an odd workplace view, but in my case it fits.

I'm a molecular geneticist—genomics, gene expression, cloning, and the rest of that good stuff – and these little guys are some of the fundamentals of what I study. In many ways, my field is actually about the flow of information in genes; how a code is represented in that mother of all blueprints and gets read to construct something so detailed and nuanced as life. My area of interest is how the information in that chain is used and communicated. It almost always happens in the same way; DNA to RNA to protein. It's as good a slogan as any, and from time to time we even get to call it dogma.

More important than this dogma, is the way my field appears to me to be so much bigger than the molecules I study. Molecular genetics represents some of the most exciting, profound, communal, and frightening aspects of the collective scientific endeavor. Its speed of advancement defies belief, and its effects on the social, cultural, political and economical issues of the day do not afford the luxury of ignorance.

That's why I sit at my desk and look at that DNA; to remind myself of the larger importance of those molecules on my window not only to myself, but to everyone else. I see that I am a participant in a greater flow of information—from expert to layman, from creating the trenches where research happens to leading the tours that engage our local community.

I suppose this isn't a fashionable reason to do science. Perhaps a more proper reason is to talk of the glory and honor of being "first" —the first to discover, to see, to understand. But in my mind, that privilege is severely limited to just one or a few. Frankly, I have my sights on something bigger: a privilege that can be shared with as many people as possible; to make science come alive.

Scientist to citizen to decisions made – wouldn't that make a lovely dogma as well?

THE VON TRAPP CHILDREN SPEAK TO A GENETICIST

LIESL: Why is it that we can all sing very well?

GENETICIST: Liesl, that is an excellent question! And essentially one that boils down to the classic debate of nature versus nurture. Are your genes responsible for this particular talent, or has it more to do with your upbringing? Looking at this scenario objectively, I would have to say that it is both. There have been reports that the ability to have perfect pitch—that is the ability to distinguish musical notes without points of reference—is a hereditary phenomenon, thereby strongly suggesting a genetic basis. This would seem to be supported by your father's musical talent as well. Of course, you've also had the benefit of being tutored by your wayward novice governess with all-world pipes, Maria.

In conclusion, like most things pertaining to our individuality, we are influenced by both our biology and our surroundings.

GRETL: I think Liesl is very beautiful. Why am I not as pretty?

GENETICIST: Assuming no mutational errors occur during the production of sperm and egg cells, there was approximately a 1-in-70,000,000,000,000 chance that you would have been an identical clone of your sister. If you included the multitude of mutational and regulatory events that ensue during this process, that statistic would escalate to an even smaller chance that is, quite frankly, unfathomable to calculate. How did I get to this absurd number? Well, one must realize that your genetic instructions are housed as a collection of 23 pairs of chromosomes (i.e., 46 in all). In other words, it is correct to say that each human has two sets of instructions—one given to you by your father, and one by your mother. If you keep in mind that your parents themselves also have 23 pairs of chromosomes, and you realize that the child may receive only one from each pair, the likelihood of siblings having the same 46 chromosomes is the fantastic number mentioned above.

However, Gretl, do not fret. You are the youngest of the lot and still have a good chance to blossom into a stunning flower like your sister Liesel. Furthermore, cosmetic surgery these days I hear is quite impressive. And then there is always the chance of Liesl having a disfiguring accident—I hear she may be a Nazi sympathizer, which is never a good thing.

FRIEDRICH: Yes, Liesl is hot. Sometimes, even I have feelings for her. Why is it bad for me to feel that way?

GENETICIST: Incestuous relationships, as well as being frowned upon by most of society, are also disadvantageous from a biological point of view. In the genetic world, diversity breeds fitness. One example is to imagine the following. You have a set of genes that determine the ability of your immune system to recognize and combat various pathogens. Your sister Liesl also has a set of genes that do the same thing. And because you and your sister come from the same genetic pool (you have the same parents), Liesl's immunity is quite likely to be similar to yours. Do you not see that the net effect of this is that you would create offspring with a limited repertoire of immune-system genes? Compare that to your having a child with, say, Marcia from *The Brady Bunch*, and you will note that this union will create offspring that have the benefit of a wider genetic pool (your parents and Marcia's parents), thereby allowing your children to acquire a more diverse and fitter immune system.

Also, dude, she's your sister.

BRIGITTA: Why do all of our siblings have blondish hair and blue eyes, whereas Marta and I have dark hair and dark eyes?

GENETICIST: You are thinking, perhaps, that your mother was a whore? It is true that the disparity in your outward appearances is a mite unusual. However, there is no reason to believe that any adultery has occurred. Here is the reason why. Although it is generally thought—though not confirmed—that extreme blondness (as in the case of Louisa and Friedrich) has a recessive distribution, there are numerous factors that can account for your instances of dark hair and dark eyes. First, hair and eye color are very subjective terms. Is Greta or Kurt blond, dirty blond, or strawberry blond? Genetic characterization is very difficult when the observational characterization is less than strict. Second, the pigmentation of hair is normally attributed to melanin levels, which have been shown to vary greatly during different stages of a person's life. You may have noticed, for example, that a person's childhood hair color tends to be lighter than their adult hair color. Third, the amount of melanin that an individual produces is influenced in part by their environment. For instance, melanin acts to protect the person from the damaging effects of the sun's radiation. In conclusion, I do not feel that there is anything to worry about. Besides, you did not mention Liesl, who herself has dark hair. Did you omit her because you are secretly jealous of her hotness?

KURT: I think I might want to be with another boy. Is this to do with my DNA?

GENETICIST: Unfortunately, the answer is currently unknown. There have been numerous reports that have tried to implicate specific genetic regions to homosexual behavior, but presently those studies, although titillating, are at best only an indication that there is a hereditary factor for this type of sexual orientation. However, there is an abundance of ongoing research in this area, particularly with homosexual men. If you are interested, perhaps you could participate in the scientific process. Of course, it is important to remember that the Nazis do not dig gay people.

LOUISA: Why doesn't anybody remember who I am?

GENETICIST: Alas, it appears that this is because you are the second child. I would not be surprised if there are very few pictures of you. It is not, I assume, because your parents did not love you, but simply a facet of being born after the initial excitement and newness of parenthood has passed. This, of course, has nothing to do with genetics. In order to be taken more notice of, you could try different fashions, or perhaps a new haircut. In truth, Liesl could probably give you better advice, as I am, sadly, only a geneticist.

ANALYSES OF THE SIX DEGREES OF SEPARATION OF BACONS OTHER THAN KEVIN BACON

Sir Francis Bacon, British philosopher, essayist, and scientific revolution advocate (1561 – 1626):
Quite a few of them are dead.

B.L.T., sandwich:
A lot depends on whether the lettuce and tomato count as one degree.

Bacon County, Georgia:
Geographically speaking, could get you as far as Florida or South Carolina

Canadian Bacon, meat cut:
Network probably not as good as Kevin Bacon's, unless of course you're referring to pigs.

Roger Bacon, Franciscan friar, English philosopher, and one of the earliest advocates of the scientific method (1214 – 1294).
Sadly, all dead.

ISAAC NEWTON, STANDING ON THE SHOULDERS OF GIANTS. EXCERPTS FROM HIS DIARY

"If I have seen a little further it is by standing on the shoulders of Giants."
Isaac Newton in a letter to his rival Robert Hooke, 1676

- – -

May 14th, 1665
Went to the post office today, thinking that I'd be picking up a grant proposal from the Royal Society. Imagine my surprise when I turned up and instead of a *grant*, there was a *giant* waiting for me.

May 22nd, 1665
A week later and I'm still a bit confused on what to do with the giant, especially since it follows me relentlessly. Friends have not been much help in this regard; enemies even less so. Some have even foolishly suggested my kicking it in the shins or standing on his shoulders.

May 22nd, 1665
Do not kick a giant in the shins. Ever.

May 27th, 1665
Stood on the giant's shoulders today. Surprised to say that it was wonderful. The world looks so different from this new vantage point, and my head is spinning from new perspectives.

May 28th, 1665
Have decided that I am never coming down.

June 19th, 1665
My cat got stuck in the apple tree today. Luckily, I can easily reach whilst on the shoulders of the giant.

June 24th, 1665
Cat got stuck in the apple tree again.

July 17th, 1665
Let's call this a lesson learned in unintended consequences. In essence, a few weeks ago, I was all pleased with myself since I had just invented the cat door – but you know what? Turns out, this was not a good idea. Stupid cat is now letting itself out and getting stuck in the apple tree daily now. The giant, fed up, has left.

August 27th, 1665
Picked up a new giant today. This one is Welsh and not averse to cats.

September 16th, 1665
With the fall upon us, we find ourselves very popular amongst all the apple tree owners. Our height makes us excellent and efficient harvesters. Indeed, I feel a bit like a celebrity, albeit a celebrity paid in bushels of apples.

October 4th, 1665
The giant and I are making apple sauce. This is actually quite difficult when standing on a giant's shoulders.

October 11th, 1665
More apple picking today! More apple bushels in my kitchen!

November 2nd, 1665
I swear if I ever see another apple, I will fucking kill someone.

December 16th, 1665
The giant and I had a grand time at our first Christmas party. He had fun dressing up as Father Christmas, but it was kind of weird when all my friends wanted to pretend to be little and sit on his lap.

December 19th, 1665
Back from my ninth Christmas party. The giant and I are really popular!

December 21th, 1665
Just had the horrid realization that we have only been invited to these so called "Christmas parties," on account of our height. Turns out we are useful for putting star and angel ornaments on the top of really big Christmas trees! I feel so used.

February 10th, 1666
Now, I am starting to get annoyed by the many *many* locals who constantly

come by and ask for some sort of giant related help. Dusting off ceiling cobwebs, hanging up large paintings, and reaching for books on the high shelf – it all gets a little old after a while.

March 28th, 1666
The giant has accidentally stepped on the cat. This seems to be bittersweet.

March 30th, 1666
The weather is starting to clear a bit, but the giant seems different. He seems melancholy and distant. The sadness is especially noticeable when I am standing on his shoulders, as they tend to be hunched these days.

June 6th, 1666
I brought home a new cat today, but the giant seemed not to notice. I am genuinely worried. Maybe I should get off his shoulders? But then again, I don't want to act too hastily.

September 3rd, 1666
It is fall again. After a difficult few months, the giant has decided to leave. In a strange way, being back on solid ground feels right. Even the apple trees look pretty again. Maybe, I'll even try sitting under one tomorrow…

CARTOON EPISODES ABOUT SCIENCE

***Peanuts* episode: "That's Biotechnology, Charlie Brown!"**

Charlie Brown loses yet another kite within the branches of his nemesis, the kite-eating tree. However, Linus cleverly observes that this action is not unlike the concept of *phytoremediation*—whereby green plants are capable of removing pollutants from the environment. Linus, along with Sally as his doting lab assistant, immediately sets upon cloning this particular tree, and goes on to secure a patent for "the use of the kite-eating tree to remove kites and other airborne contaminants from the air." As a result, Charlie Brown, Linus, and Sally embark on a biotechnology business venture that quickly makes them extremely wealthy. Empowered with his new affluence, Charlie Brown finally tells Lucy to "fuck off."

***The Super Friends* episode: "Wonder vs. Wonder"**

When it becomes clear that a mission is botched because Wonder Woman is clearly visible in her invisible jet, unhappy murmurs begin to surface within the Super Friends' organization. In particular, Zan, of the Wonder Twins, is merciless in his teasing of Wonder Woman. It also doesn't help that Wonder Woman, herself, is generally not impressed with his otherwise useless superpower ("Form of a bucket of water? What in Amazon is that about?")

In any event, Batman decides to put his scientific mind to work by fixing the jet and soon discovers a small error in the optics of one of the twenty cameras that are responsible for the illusion. Unfortunately, this only seems to encourage Zan further, who torments Wonder Woman on the seemingly mundane manner that invisibility is conferred. ("It's literally all done with cameras! What a loser plane!") In the end, fed up with Zan's abuse, Wonder Woman soundly beats the crap out of him.

***The Simpsons* episode: "My Fat Bonehead"**

Guest starring as herself, Jessica Simpson visits Springfield to teach Homer the ropes of becoming a southern gentleman (à la My Fair Lady). This goes as well as expected, and Bart in particular becomes completely smitten by the young lady. However, it is then revealed that Lisa is recently diagnosed with acute myeloid leukemia, and furthermore is in need of a bone marrow transplant. Miraculously, Jessica Simpson is the perfect match, which culminates in the use of genetic testing techniques to show that she is, indeed, Homer and Marge's long-lost lovechild. Bart then has to deal with conflicting feelings of lust and the heebie-jeebies from this apparently incestuous crush.

***Dora the Explorer* episode: "¡Hola! I Have a Brain Tumor!"**

In this episode, Dora visits her doctor to complain about her dry, red, and itchy eyes. The doctor quickly solves the problem by advising Dora to try blinking for a change. However, at this visit, the doctor quickly suspects Dora is plagued with a more serious psychosomatic condition, since she continually refers to a talking backpack, a talking map, and a talking monkey with a perceived preference for sturdy yet red colored footwear. When Dora continues to stare off into the distance and ask bizarre and loud questions towards no one in particular ("What was YOUR favorite part of the day?"), the doctor decides to take matters into his own hand and schedules her for a CAT scan.

ON GENETICS, RADIOHEAD, AND THE PLIGHT OF KID A

Lately, I've been listening to a lot of Radiohead.

This is not so surprising, because in general I would have classified myself as always being a fan. I find their musical compositions, instrumentations, and Thom Yorke's heart wrenching vocals a compelling mix: although admittedly, I do sometimes miss the days when some of their material was a little less challenging to radio ears. Still, truth be told, I'm such an admirer that I even use Radiohead in my role as a science educator.

Specifically, I might play one of their stranger sounding songs and then project on the wall my students with the following statement:

"The song you are listening to is the title track of a Billboard number one CD. The song and CD appear to have been specifically written and dedicated to the first putative human clone."

Then I'll ask my students, "True or false?"

Inevitably, the students will concentrate on the song coming over the speakers, and they will note that what they are hearing, sounds somewhat unconventional. It's distorted in many places, and doesn't adhere to any semblance of common chord progressions. It is dissonant in nature and definitely does not sound like an obvious reflection of a Billboard number one CD.

And then the reveal: yes, the statement is true. It turns out that the song is the title track of the CD *KidA* which on October 21st, 2000, did chart at number one. Not on the Alternative Billboard chart, or the Electronic Billboard chart, but on the whole-freaking-thing Billboard chart. Furthermore, it does apparently connect with the topic of cloning. At least, that is what you might surmise when you read what Thom Yorke once said on the subject:

POSTED BY Thom ON JULY 30, 2000 AT 23.39:21:
IN REPLY TO: Thom, why Kid A?

dedicated to the first human clone.
i bet it has already happened.

In any event, I tend to use this little activity as an excuse to talk about science perceptions, and how topics like genetics can be quite pervasive in popular culture. Not only pervasive, but often in a manner that denotes a negative connotation. Think of all the movies where genetics plays an integral part of the conflict in the plot; be it biological warfare in spy franchises, clone armies in science fiction sagas, or even battle scenes between Pokemon characters and their angry identical copies (Yes, there is a central cloning plot to the movie, *Mewtwo Strikes Back*).

In its own way, the song *Kid A* also presents a certain negativity in its tone that is both foreign sounding, uncomfortable, and haunting. Maybe Radiohead did this to add their own criticism of technology to the mix, or maybe it's more to do with an artistic representation of how that clone might feel forced through such a strange incident. Of course, maybe the song just came out that way because the band thought it sounded cool.

Regardless, all of these different ideas also bring up a strange thought I've been considering.

That is: let us assume that Radiohead were right - that they were right about the first putative human clone being produced, or say, conceived roughly in the year 2000. This would mean that their sonic ode would have been temporally, fittingly, and perhaps even devastatingly relevant. Furthermore, it would mean that somewhere out there, on our busy little planet, is a child who is technically the song's point of inspiration, and also a child who is now probably around 12 years old.

What if this child found out? What would it think when it learns that an iconic British band wrote a song for him/her. Would the child have questions to ask? Would the child's parent (presumably the supplier of the DNA) suggest he/she write a letter? What would that letter say?

Perhaps, he/she would comment on the science?

"To be honest, I'm not even sure what "human cloning" means, since right now, the science I'm learning at school, is mainly to do with habitats, and food chains and why the sun is super important. Someone did once try to explain the cloning thing to me, but I just got confused."

Or maybe there would be mention of Radiohead's popularity:

"My parent tells me that you are famous. In fact, they tell me that you are really famous. I have to admit that this is surprising to me, since I have never even heard of you, or even seen you on TV. As well, I tried to listen to this KidA song, but I think the song was broken since it sounded really strange..."

Which, not surprisingly, might culminate in a request to meet, or possibly something more selfish:

"I'm wondering that since you are super famous musicians, that maybe you can introduce me to some of your friends. For instance, have you heard of Taylor Swift? My friends keep playing her CD – it's really good. It would also be cool to meet Katy Perry or Jay-Z..."

Which leaves us with an idea that is so deliciously meta: what kind of music would the *real* Kid A actually listen to? From my own experience, having kids of my own, I'm pretty sure that this 12 year old Kid A will not be listening to Radiohead. Which is actually something that makes me feel sad. There's a loss of tidy symmetry there - although the glimmer of hope is that maybe it happens to feel sad enough for Radiohead to write a new song about.

HUMAN GENE COMMONLY ASSOCIATED WITH CANCER OR DROID FROM STAR WARS?

1. p21RAS
2. C-3PO
3. CD45
4. p53
5. C-SRC
6. RIC-920
7. FOS-JUN
8. R2-D2
9. 8D8
10. C-MYC

1, 3, 4, 5, 7, 10 are oncogenes: 2, 6, 8, 9 are droids from Star Wars.

THE NEW SCIENTIFIC METHOD

Make an observation.

Take a photo of it with your phone. Apply cool looking image filter, tweak with selective blurring, and then share via Facebook, Twitter, Instagram, your blog, etc.

Provide a trite but punchy comment that explains your observation. This is your hypothesis. OMG!

Wait for comments. This begins peer review.

Break time! Catch up on celebrity gossip!

Keep waiting for comments. Don't take lack of interest personally. Try to feel better by repeating to yourself, "I do this because science is important, I do this because science is important…"

Retake photo, or better yet, record a video. This time, make sure a funny cat happens to be in the frame of view. Alternatively, modify your hypothesis by including weak references to pithy teen pop stars of your choice.

Post this video hypothesis on YouTube. Hope for it to go viral. Under no circumstance, do you acknowledge the fact that the word "viral" originally referred to a biological concept and not to computers.

Break time! Catch up on celebrity gossip!

You get comments! You feel validated, and decide to upgrade your phone and/or possibly adopt a cat. This is your technology investment. LOL!

Skim comments, paying attention only to comments from people with attractive profile pictures. This is your data.

Apparently, data is very confusing and occasionally rude. WTF? Nod knowingly to yourself that research is hard, frustratingly incremental, but important nevertheless. Make animated gif of yourself nodding knowingly. Post this on Tumblr.

Break time! Catch up on celebrity gossip!

Data is still confusing? #facepalm Attempt to re-assess ideas by consulting transcriptions from the likes of Dr. Oz, Jenny McCarthy, and/or your local politician. Consider this expert peer review.

And if all else fails, try googling.

Break time! Catch up on celebrity gossip!

Repeat until consensus is formed. FTW!

ASSORTED RAYS: RANKED ACCORDING TO COOLNESS

6.
Ray Romano
Is it just me or is this guy too funny? I mean, that thing he does with his TV mom and wife just cracks me up. Plus, he once made $50 million bucks in one season, which is totally cool, and is in no way the reason for putting him on this list. Too bad about the TV kid twins, though – I mean, what's up with their foreheads being so massive? It doesn't look natural.

5.
Cosmic Rays
These are the rays that gave the Fantastic Four their powers. But even cooler – in astrophysics, they are basically high-energy outer space particles that make their way to the Earth. How awesome is that! It's like they're all around all the time. Plus, I did some reading on them and found out that the most energetic recorded was 10^{20} eV! I don't even know what an eV is, but its got to be pretty cool. Also, 10^{20} is one big number – that's a one with 20 zeros behind it. Once in my car, I even tried counting to it, but only made it to 214. I think I could have made it all the way but True by Spandau Ballet came on the radio and I hate that song.

4.
Ray Bans
These sunglasses are as cool as it gets. Unfortunately, I already wear prescription glasses and I'm too cheap to get prescription sunglasses. This means that when I put on a pair of Ray Bans, I either have to put them right on top of my prescription glasses, or alternatively I take my prescription glasses off first, put the Ray Bans on, and then put my prescription glasses on top of them (I can't see without them). Anyway, I don't think this is how Ray Ban intended it, but I suppose this is why it's not at the number one spot.

3.
Stingrays
How awesome are these fish? They swim with those kick ass pectoral fins and have a nasty serrated sting that – get this – is coated with toxic venom.

I'm a bit surprised there's no TV show for them – you know, like Flipper, except when you piss it off, it might kill you. I mean, if that's not prime time then I don't know what is. Or it could even be like a comedy because, did you know a Stingray's eyes are on the top side and its mouth is on the bottom side? That's right, people; the poor fish can't see what it's eating! Man, that kind of comedy just writes itself.

2

The "Re" in "Do Re Mi"

O.K. so not technically a "ray" – but this one rocks! First, am I the only one who thinks Julie Andrews was pretty hot back then? More so, when you realize that she's playing the guitar for real in the movie – double score! On top of that, there the whole "drop of golden sun" line, which I'm guessing is in reference to that whole quantum physics wave-versus-particle thing – it's a shame that whole subplot was edited out of the movie. Some nuclear explosions would have really taken that movie to whole different level.

1.

(Tie) Gamma Rays and X-Rays

I've decided that these two rays are tied for first place, because you know what? Sometimes, in physics land, they are actually the same thing! Although, you could probably care less since maybe that, in itself, is not that cool. But how awesome are x-rays? You can see your teeth and bones, for Christ's sake! Plus, you have to wear lead aprons when you work with the stuff, and nothing says "attractive" like a kicking lead apron. And gamma rays – did you know these are the babies that gave us the Incredible Hulk? Although what's up with his shirts always ripping to shreds and his pants always staying together? I thought your gluteus muscles are supposed to be the biggest in the human body. Anyway, I'm not actually complaining – it's not like the Hulk is hot like Julie Andrews or anything, although he does look like the sort of dude who would also have a problem with Spandau Ballet.

AM I EVERYWHERE?

Last year, I had a decidedly Jungian experience, which is odd for me as a rational scientific sort. This synchronicity event happened at a local bookstore where I was searching for an issue of a Canadian magazine called *Maisonneuve*, a publication nice enough to print an essay of mine. As I was looking over the rack, my eyes wandered and noticed an issue of *The Believer*[1], where lo and behold, I saw my name on the front cover – a very observable and clear "by DAVID NG", written with agreeable font, and even flanked by two pretty star icons. I hurriedly flipped through the magazine to see if I could find any information on this author[2], confused that my life had perhaps become so busy that I was submitting articles without even knowing it.

Moments later, I found my copy of *Maisonneuve* and flushed with my first real experience in publishing, looked upon its front cover. This might seem silly but this otherwise personally important moment came with a small pang of disappointment. Although the title of my article had made the front cover, my name didn't. Somehow, David Ng had stolen my thunder. Or was it more accurate to say that I (David Ng), stole my own thunder? Anyway, it did make me think a bit.

It also compelled me to look deeper. That and the fact that the Believer piece was simply entitled "v4.0" which lent a deterministic feel to my pursuit (how many versions of me are there)? Indeed, my name was already fraught with a few unconventional inconveniences – my surname in particular. It has, for instance, no vowels which growing up in England was an oddity that frequently confounded my grammar school teachers. Here, the phonetic pronunciation of 'Ng' often took the form of a caveman-like

---

[1] Maisonneuve issue no. 9, June/July 2004 and The Believer issue no. 13, May 2004.

[2] In a surreal twist of fate, since the time of writing, the author of the Believer piece and I have actually exchanged a few emails. He sounds nice, and I've even asked if he was intereste

[2] In a surreal twist of fate, since the time of writing, the author of the Believer piece and I have actually exchanged a few emails. He sounds nice, and I've even asked if he was interested in collaborating.

uttering. Of late, my surname has also made naming my children an activity fraught with caution. I couldn't name my son "Bob", "Dick" or even "Jack."[3] It would simply raise the 'getting beat up during school' quotient far too high.

But in this case, seeing my alter ego in print, elicited a more guttural response. As if my very essence had been trespassed upon. Almost, as if I had been cloned and not told about it. An interesting predicament no doubt and so, with the pervasiveness of media and the possibilities provided by science and technology, I was curious to see how a person's sense of self, their very individuality, and even their fragile ego could take a beating when their name *exists* in other contexts[4].

I started with the local phone book, and found fifteen other David Ng's. Sixteen if you count my own unlisted information, twenty if you counted the Dave Ng's, and possibly many more that don't make their way to the white pages. This meant that if I was wandering the streets of Greater Vancouver, I would have at least a one in 125,000 chance of meeting another David Ng[5], odds that are happily much better (and hopefully more enjoyable) than the yearly one in 800,000 chance of getting struck by lightning. Whilst looking through these addresses, I also noticed that one of them was listed as a certified general accountant and another as a doctor, a surgeon in fact. Perhaps one of these days, I might be so lucky as to stumble upon the elusive "Reserved for Dr. David Ng" parking stall. Free parking is always a good thing[6].

---

[3] For some reason my friends laughed most at the prospect of naming our child "Isaac" which would be pronounced, no proclaimed "I'sa King!" For the record, my children's names are Hannah and Ben.

[4] This whole idea of uniqueness associated with a person's name lends credence to one of my theories about why certain celebrities might name their child "Apple." The other hypothesis, of course, is that they lost a bet.

[5] According to www.whitepages.com, at the time of writing, there were only 87 David Ng's in the United States. This is interesting because the same website was able to pick out most of the 15 or so David Ng's in Vancouver, Canada, thereby giving credibility to its searching prowess. Is it possible that I happened to be living in the David Ng mecca of the world? We should start a fan club.

[6] I have actually on one occasion parked in a stall reserved for "Dr. Ng" Ph.D.s are useful sometimes.

Searching the web was where my vanity took the biggest hit. At the time of writing, a google.com search for 'david ng' yielded 2,290,000 hits, found in no less than 0.24 seconds. Using yahoo.com, a speedy 0.17 seconds was all that was needed to find 841,000 results. Overall, the top ten results for each of these search engines yielded some pretty insightful results and included in no particular order:

two writers (neither of which were me)
one movie star
one economics professor
one link to find David Ng at Amazon.com
one reference to William Hung of American Idol fame
one animatronic head
one anime artist
one rock collector who works with alpha, beta and gamma radiation

Personally, I ranked in at #14 with google.com, but did not even rank in the top 200 with yahoo.com. I suppose I should be happy to have at least made the top 20[7] with one of the major search engine.

Because of my biology background, I next decided to look for myself in an even bigger context. In nature, itself, if you'd like. You see, my name is actually one that fits the FASTA system, a language of letters used by biologists in writing out our own and every other organism's genetic code[8]. Could D-a-v-i-d-N-g be found in the very fabric of our DNA, the gene products that build us, that move us, that control us? Apparently yes.

Currently, the code for DAVIDNG[9] can be found only in the genetic instructions of one Thermus Aquaticus, a very old species of bacteria that

[7] On the other hand, I rank no. 3 with google and no. 7 with yahoo when searching for 'sciencegeek.' Not sure if this is a good thing. Also note that this was back in days when personalized search did not exist.

[8] To be more precise, D-a-v-i-d-N-g can be used to look for protein sequences. Proteins are the things that DNA ultimately codes for, and Atkins aside are the things that do all the interesting things in our body.

[9] Or aspartic acid-alanine-valine-isoleucine-aspartic acid-glutamine-gylcine. In all, proteins are composed of twenty different amino acids, which each have a letter designation. Basically, as long as your name doesn't have a 'B','J','O','U','X' or an 'Z' you may exist in the genome of something or another. To do this, you can go

have the nifty ability to grow in boiling hot environments, thereby making them an unfavorable pet choice for children. To be honest, this was actually pleasing to me, to know that DAVIDNG wasn't literally everywhere in all manner of organisms. By contrast, the code for ELVIS is very common[10]. Unfortunately, my own curiosity got the better of me and I also took it upon myself to check if DAVENG[11] was present in the various genomes of various organisms. Turns out, in a major knock against my individuality, DAVENG was everywhere. In fact, it can be found in the genetic code for:

Many bacteria (such as e.coli, mycobacterium, helicobacter to name a few)
Ustilago Maydis (a commonly studied agricultural fungus)

Arabidopsis (currently a favorite model system for plant research in the world)

Houseflies (currently a favorite model system for invertebrate research in the world)

Mice (currently a favorite model system for mammal research in the world)[12]

Needless to say, it's a good bet that when you come down with a fever, sniff a flower, kill an annoying fly, or chase a furry rodent away, the organism you are dealing with is subtly characterize with a DAVENG in its very being. Apparently, I *am* everywhere.

Which leads to a silly final thought that I only think of because of my background. In some twisted respect, here is a situation where my previous

---

tohttp://www.ncbi.nlm.nih.gov/BLAST/ and click on the 'search for short, nearly exact matches' in the PROTEIN subheading. In the new page, enter your name, and then hit the 'BLAST' button. Results are as they were at time of writing.

[10] Or glutamic acid-leucine-valine-isoleucine-serine. Curiously, you can also look for ELVISISDEAD and ELVISLIVES in an attempt to rest this tired question. Unfortunately, neither of these sequence has a direct match, and so the mystery remains unsolved. I should add, in case you are the worrying sort, that there is absolutely no chance of GEORGEBUSH, CELINEDION, or ELMO occurring in the human genome.

[11] Or aspartic acid-alanine-valine-glutamic acid-gluamine-glycine. Isn't science fun?

[12] No matches in the human genome, the closest being a 'DAVIDN.'

offhand comment about having a clone becomes intriguing. Even more so, when one considers that I happen to be an individual who *theoretically* has the technical ability and the equipment access to pull it off[13].

In the end, I would need to ask whether the world really needs more David Ng's? Another one that has been created to compete with the multitude that are already out there? And ironic in that in the end, it would probably have both a virtuous and deleterious effect. Virtuous in that my own personal notoriety would surely rise, thus ensuring my own prominence in this land of David Ng's. And yet deleterious because there is a chance, arguably a very good chance, that it would be the clone itself who attains the top spot.

---

13 Surprisingly and frighteningly, it doesn't take much. Any American infertility clinic probably has the equipment, and technical ability to do it. In Vancouver, I know of at least two robotic microinjectors, many tissue culture facilities, and over a million wombs. It is, however, a very tricky procedure best done before coffee.

A SCIENTIFIC PROPOSAL TO THE EXECUTIVE PROGRAM DIRECTORS OF THE ABC, CBS, FOX, AND NBC NETWORKS

Dear Sirs,

I know an omen when I see one, and it needn't even involve a two-headed goat. As a scientist with a background in cancer research, the revelation I'm referring to is a bit of homework I did on the average yearly amount of money spent on programming by your television networks (about $1.5 billion). A number which strangely mirrors the average amount of money given last year to each of the 18 institutes within the National Institutes of Health, an organization that is the U.S.'s backbone of publicly driven medical research. Clearly, this is a call to merge the two enterprises together. So in the interest of public health, and given the pervasiveness of reality TV, I wish to expound to you four possible examples that demonstrate the feasibility of this union.

Real Science, Real People:

In the early 90s, studies were conducted whereby a single male mouse was presented with a plethora of different female mice. What was discovered was that the most desirable females had immune system genes that were most *distinct* from the male suitor. In other words, the female picked had a particular genetic background. Such a mechanism of mate selection would please Darwin since the offspring produced would inadvertently benefit from the most diverse, or most advantageous, immune system. More pressing, however, is the question of whether this decidedly unromantic notion pertains to mate choice in humans?

Fortunately, we can now answer this question by asking the participants of programs like *The Bachelor* or *The Bachelorette* to provide a blood sample along with their video profile. This way, research can finally circumvent the sticky ethics of conducting such experiments on humans. On the plus side, this research opportunity should also generate its own built-in funding infrastructure as it can be easily applied to beat Vegas odds.

Save Money:

Currently, every drug used for medicinal purposes in the United States needs to navigate through the strict and often precarious guidelines imposed by the Food and Drug Administration. This is an extremely long and expensive process, averaging 15 years and upward of $300 million in financing from discovery to product. Inevitably, most of this arduous process is due to the proper design and delivery of human clinical trials that examine drug efficacy and safety. Why not incorporate these trials into television shows like *Fear Factor* or *Survivor*? If contestants are willing to drink the seminal fluids of cattle or eat squirming maggots the size of your thumb, wouldn't these same individuals revel in an opportunity to eat untested drugs? We could even have a "totally untested" and a safer "well, the mice survived" version of the same contest! In any event, millions of dollars would be saved.

Promote Technology Development:

Medical research is largely driven these days by the ingenious design of equipment that can do new things or do old things better, faster, bigger, cheaper, and safer. This to me is an invitation to incorporate medical technology development into reality TV. Why can't *Junkyard Wars* showcase a competition to build the fastest DNA sequencer. Or viewers watch an episode of *BattleBots* that pits equipment used for insulin production. If *Extreme Home Makeover* can build a whole new environment in seven days, then why can't you "fix that genetic mutation" in the same seven days. It's no surprise that ingenuity often percolates under tough situations, and I can think of no tougher than a scenario where contestants only have 48 hours and a $1000 budget to meet their objective.

Fostering Interest in Science Careers:

If we can have programming that features Donald Trump searching for a skilled apprentice, why can't we use the same template to attract top graduate students. It should be simple enough to invite a feisty Nobel

Laureate with an ego big enough to oversee the process. Just think of the entertainment value generated by having a team of young researchers told "Your project is to work together and come up with a cure for cancer *in three days*. And don't forget—if you fail, you will meet me in the seminar room where somebody will be fired!" I mean, really—this stuff sells itself!

To conclude, I hope these four simple examples illustrate the opportunity at stake. It would be a great shame to not utilize these two great charges for the benefit of all. Now if we can only get the Food Channel on board—maybe an episode of *The Iron Chef* with two-headed goats as the special ingredient?

Sincerely,
Dr. David Ng

MOTHER GOOSE AND THE SCIENTIFIC PEER REVIEW PROCESS

Jack and Jill went up the hill.
To fetch a pail of water.
Jack fell down and broke his crown.
And Jill came tumbling after.

First of all, we are not sure there's enough clarity in this text. Scientific literature, in particular, should leave little room for confusion. Where exactly did Jack fall down? Into the well? A little ways down the hill? All the way down the hill? It's just too vague. Worst still, we're not convinced that the science conducted is of high enough caliber. I mean really, who would be stupid enough to put a well on the top of a hill? In conclusion, we feel that this manuscript should be rejected in its current state, but are not opposed to viewing a revised version in the near future.

—

Twinkle twinkle little star.
How I wonder what you are.
Up above the sky so high.
Like a diamond in the sky.
Twinkle twinkle little star.
How I wonder what you are.

Initially, we were quite intrigued by your work, especially since it appeared to contain several elements that merit genuine excitement. However, it was then brought to our attention that this body of work had remarkable similarities to a previously published report (The Alphabet Song). It was upon further investigation, that our worst fear was confirmed to be true – that this manuscript constitutes an act of plagiarism. We must state that we feel this to be a serious breach of scientific ethics, and must therefore strongly decline your manuscript.

—

Humpty Dumpty sat on a wall.
Humpty Dumpty had a great fall.
All the King's horses and all the King's men.
Couldn't put Humpty together again.

Although otherwise promising, the reviewers felt that the research in its current state is incomplete. Quite frankly, it was agreed that your principle subject needed to be put back together again. Several of the reviewers suggested courting the expertise of a mathematician who could perhaps create an appropriate algorithm to solve this problem. Alternatively, one reviewer suggested glue. As a final note, questions were also raised regarding the treatment and well being of Mr. Dumpty. Why exactly was he made to sit on the wall? And why exactly would you allow horses (of all things) to put him together again. No matter, the reviewers overall impression was that if you were able to address each and every one of these issues, they would see no problem entertaining a revised version.

—

Hey diddle diddle, the cat and the fiddle.
The cow jumped over the moon.
The little dog laughed, to see such a sight.
And the dish ran away with the spoon.

The reviewers felt that not enough data was presented to support your claims. For example – how many times did your group observe the cow jumping over the moon? From the text and supporting figures, it would appear that you base this conclusion on one data point as no calculations regarding standard deviations were presented. As an analytical journal of high repute, the reviewers felt that this is simply not acceptable. In addition, several of the reviewers felt that the word 'diddle' was inappropriate, and should have been replaced by the more scientifically correct, 'Hey fornicate fornicate." Because of these, and other problems, we are sorry to inform you that your manuscript has not been accepted for publication.

—

Rub a dub dub, three men in a tub.
And who do you think they'd be?

The butcher, the baker, the candlestick maker.
Turn'em out, knaves all three.

Thank you most kindly for allowing us to see this marvelous manuscript. We feel that it is a great privilege that you and your colleagues decided to submit it to our journal. We truly feel that it represents seminal work that could even one day lead to a Nobel prize. To be frank, we were quite surprised to receive your submission, in that we all felt it could have easily been accepted by the more high profile publications (The Nature and Science journals for instance). In any event, we are very pleased to inform you that, we, the reviewers are unanimous in our decision to accept your manuscript.

CHAPTER TITLES FROM MY CREATIONIST TEXTBOOK

Thursday
Saturday
Embryos
Friday
Homosexuality
Wednesday
Monday
Thermodynamics
Tuesday
Candle Making

AN OPEN LETTER TO 0.7%

Dear 0.7%,

Let me say that it is an honour to make your acquaintance! Congratulations on having been chosen as the target percentage of GDP rich countries should send in foreign aid to support developing nations. You should be extremely proud of yourself, having beaten out 0.6 and 0.8 by the slimmest of margins—just 0.1. And 0.8 did a very strong audition, performing a number—the musical kind—from *The Lion King*. But your rendition of "SexyBack" captured the right mood.

Using you as a benchmark for foreign aid is brilliant—what could be easier than wealthy nations contributing 0.7% of their gross domestic product to alleviate the problems of the Third World? But the cold, hard fact, my friend, is that nobody is taking 0.7% seriously. So let me tell you what we have in mind, branding-wise.

For starters, there is the whole math thing. It's a buzz kill. Who can multiply a number by 0.7 on the fly? And—no offence—you're not easy to work with, like 10 or 2. Or 100! 100 is golden. But would you read a book called *0.7 Years of Solitude*? And, technically, you're like, 1/100 of 0.7—you're a series of calculations that can stop a good-hearted impulse in its tracks:

"*Hey, our GDP is lookin' good. Let's flip Somalia something for digging wells or whatever . . . let's see, 0.7% of $68 billion is . . . (government minister counts on fingers of both hands for thirteen or fourteen minutes) . . . oh, never mind. Let's go with the guy who sent us those wildlife calendars instead . . .*"

The other thing is, although you're offbeat you're not good-quirky. Ask anyone to pick out an exciting number, and nobody will say "zero-point-seven." You certainly don't have the brooding complexity of pi, the lonely mystery of zero, or the loud celebrity of "ONE MILLION!" Those kind of figures just sell themselves. You're Wilco playing to the *High School Musical* crowd.

And what about longevity? I know you've been around since 1969, with your big-gun, United Nations birth cred, and that you were part of the UN Millennium Development Goals, a show that had some legs. You were tight with Lester Pearson. Still, it's hard to see you as a franchise:
Return of the 0.7
1.7
2 x 0.7
0.8

Do you see what I'm getting at? They all suck. And they'd still suck even if you could get Will Ferrell to play 0.7 as conjoined twins, 0.35 and 0.35. Even taking 10.3 guys out of the cast of *Ocean's Eleven*. It doesn't matter—the sequel titles just kill it. The number 7 does have some pedigree, so it's not necessarily the fault of your digits.

Nobody has your digits, and I mean that. The trick is to fall on the good side of the marketing equation. Here, you can minimize aligning yourself with, for instance, deadly sins and years it takes to itch, and, at all costs, avoid the dwarf debacle (although *Snow White and the 0.7 Dwarves* does have a kind of morbid appeal).

Instead, to promote you we'll be leaning more toward wonders of the ancient world, magnificent cowboys, handy convenience stores, and possibly even ambiguous-tasting lemony-limey soft drinks. What does 007 bring to mind? Lovely ladies and gunplay, that's what.

And 0.7%? Combined with the acronym GDP, with its slight urological overtones? I'm thinking HIV pandemics and abject poverty. Depressing. Maybe instead of being "licensed to kill," we can position you as being "licensed to save"? Whatever.

Clearly, we have to work on it. I think our best bet might be live performance—something along the lines of *Dancing with the Stars*, with lots of glitter and skin—in which you compete with some highly visible celebrities looking to add the Third World to their resumés. The idea is 0.7% as the misfit decimal, the little engine that could, just like Jennifer Hudson in *Dreamgirls*. We're gonna make them love you!

Thoughts?

Best, Bernie Gottfried
Fractional Talent Los Angeles

A CALCULATION TO SEE HOW CUPS OF COFFEE YOU WOULD NEED TO DRINK TO KILL YOU

I'm in full on marking mode right now, which also means my uptake of coffee has increased significantly. Consequently, I'm procrastinating and thinking about strange things – such as lethal doses – especially for things we scientists particularly indulge in (like coffee, alcohol and, yes – the free cookies at Departmental seminars). So let's look at the fatality of coffee drinking? And here, for the scientist, the first place to look a little deeper is the vaulted MSDS (or Material Safety Data Sheet).

For those not initiated in this lingo, MSDS are those documents that provide risk assessment and health considerations for any and all reagents, compounds, molecules, chemistries you might care to use in a laboratory setting. Of course, the most press worthy value it often provides is the "lethal dose." Which, according to wiki is described as follows:

"the median lethal dose, LD_{50} (abbreviation for "Lethal Dose, 50%"), LC_{50} (Lethal Concentration, 50%) or LCt_{50} (Lethal Concentration & Time) of a toxic substance or radiation is the dose required to kill half the members of a tested population.[14]"

Anyway, I thought it might be interesting to do some back of the envelope calculations to bring to you, some information on how many cups of coffee to avoid drinking - so as to not kill yourself.

However, this calculation is not as easy as it sounds, because there's a certain amount of kinetics that needs to be taken into consideration. So, let's first start with a few facts and figures to get the ball going.

To begin with, if we're going to focus on coffee, probably its most potent chemical component from an oral lethal dose point of view is the caffeine.

---

[14] From http://en.wikipedia.org/wiki/LD50

However, from a purely empirical perspective, it might actually be the water content that will kill you in the end. In other words, if you drink lots of coffee and plan on doing it to induce a fatality, it might be interesting to see what scenarios are necessary for that death to be caused by too much caffeine versus too much water.

In any event, here are the numbers to concern ourselves with:

1. Average weight of a human: From wiki:

"In the United States National Health and Nutrition Examination Survey, 1999-2002, the mean weight of males between 20 and 74 years of age was 191 pounds (86.6 kg, 13 st 9 lb); the mean weight of females of the same age range was 164 pounds (74.4 kg, 11 st 10 lb)[15]*"*

Let's use **80kg** as an average.

2. A single cup of coffee on average contains about 250ml of water, and about 135mg of caffeine[16].

3. Lethal dose (oral intake for a rat, which has similar metabolism – although we should note, not identical metabolism) is about **192 mg/kg** for caffeine and **90 mL/kg** for the water[17].

4. However, the other part of the equation that we need to evaluate involves rates of elimination.

"The half-life of caffeine–the time required for the body to eliminate one-half of the total amount of caffeine–varies widely among individuals according to such factors as age, liver function, pregnancy, some concurrent medications, and the level of enzymes in the liver needed for caffeine metabolism. In healthy adults, caffeine's half-life is approximately 4.9 hours.[18]*"*

And for water – this was a little harder, because water turn over rates tended to revolve around the idea of an individual not imbiding in crazy amounts of fluids. So, for the sake of our calculations, I'll go with the following piece of information:

---

[15] From http://en.wikipedia.org/wiki/Human_weight
[16] From http://www.cspinet.org/new/cafchart.htm
[17] Both values from MSDS sheets
[18] From http://en.wikipedia.org/wiki/Caffeine

"It's Not How Much You Drink, It's How Fast You Drink It! The kidneys of a healthy adult can process ***fifteen liters of water a day****! You are unlikely to suffer from water intoxication, even if you drink a lot of water, as long as you drink over time as opposed to in taking an enormous volume at one time. As a general guideline, most adults need about three quarts of fluid each day. Much of that water comes from food, so 8-12 eight ounce glasses a day is a common recommended intake. You may need more water if the weather is very warm or very dry, if you are exercising, or if you are taking certain medications. The bottom line is this: it's possible to drink too much water, but unless you are running a marathon or an infant, water intoxication is a very uncommon condition.*[19]"

O.K. so now let's do the math.

First, an oral lethal dose for an 80kg human would extrapolate to **15,360mg** of total caffeine. This technically is equivalent to the amount of caffeine absorbed from drinking 113 cups of coffee really really *really* quickly. However, the reality is that this figure would instead result in a fatality due to water intoxication since 113 cups is close to 30 litres of water.

So let's try a different tact: by focusing on a safe water ingestion figure (i.e. 15 litres per day when spread reasonably). This works out to 60 cups of coffee over a full day, or approximate one cup every 24 minutes. Anyway, this is some pretty nasty math to figure out (since it's a half life calculation with continual replenishing going on). Anyway, if you do the math, what you find is that at the end of a 24 hour period, that average body would have retained a little less than **2500mg** (this is based on some very rough back of the envelope calculations). Not even close to the 15,000 or so milligrams needed to reach the lethal dose. Presumably still not a healthy thing to do, but within the context of our LD_{50}, it sounds doable.

And the funny thing is, by the next day, that **2500mg** would have been metabolized or cleared itself and only about **50mg** of this is left behind. Which means that the net total amount of caffeine still in a person's system if he or she were to continue drinking a cup of coffee every 24 minutes for a 48 hour period is **2550mg** (2500mg + 50mg).

---

[19] From *Can You Drink Too Much Water?* http://chemistry.about.com/cs/5/f/blwaterintox.htm accessed on January 30, 2015

It turns out that your body is potentially quite capable of dealing with such a heavy coffee dosage, because that new 2550mg level becomes 53mg by the next 24hours – therefore three days of drinking a cup of coffee every 24 minutes will result in a net retention of 2553mg (2500mg + 53mg) and so on.

I haven't had a chance to extrapolate this over the full year (365 days), but I'm pretty sure that even a constant coffee drinking regime (1 cup every 24minutes for the full year) wouldn't work out to a retention amount above the lethal dose.

All to say that your body pretty much kicks ass in its remarkable metabolism. Now, it'll be interesting to maybe dig a little deeper with regards to how messed up a person gets with that base 2500mg inside them (as I'm sure the case will be). As well, not sure what the deal would be with 15 litres of expresso shots per day – that actually may just about be enough to kill you.

CONGRATULATIONS! YOUR INEFFECTUAL GENETIC TEST RESULTS HAVE ARRIVED!

Dear Client,

We at Genetic Tests for Realz are pleased to present results from careful analysis of your DNA sample. As discussed previously, data was obtained in triplicate from our state of the art laboratories. Also note that our team of moderately trained individuals was assuredly wearing white lab coats during all experimental steps. Given these excellent standards, we are pleased to affirm that with regards to your genetic makeup, each of the statements below is presented with high confidence. Pertinent pieces of information have been capitalized for full clarity.

1. You are, in no uncertain terms, GOING TO DIE. This is unfortunate, and we feel regret on having to inform you of this, but there you have it—this is what your DNA code is telling us.

2. We contend that based on our analysis, a number of medically relevant activities are strongly advised for better health outcomes: these include (i) BREATHING, (ii) EATING, and (iii) DRINKING. Note that URINATING and DEFECATING also come highly recommended.

3. Our initial scanning of your DNA strongly suggests that you have TWO ARMS and TWO LEGS. This is almost for sure—unless, of course, you don't. In which case, our staff asserts that this inaccuracy is moot, as a genetic test was obviously NOT needed to make this sort of assessment in the first place.

4. In the unlikely event that we were wrong with the whole arm and leg thing, let us placate any doubts in the quality of our services by stating that your genetic code DOES unequivocally confirm that you MOST CERTAINLY have at least one HEAD. Furthermore, this same analysis also strongly suggests the presence of a BRAIN of some sort.

5. By incorporating only the most recent and advanced findings in genomic research, your DNA sequences clearly show that you are NOT A CAT.

Note that while the evidence for this is VERYSTRONG, we cannot currently make similar statements about your overall OPINION of cats.

6. PRELIMINARY results suggest that you are exactly the sort of individual that will score HIGHLY for the ESBFGT[20] trait.

7. Given your predisposition for the ESBFGT trait, we feel it is important to inform you that we have lots of other DNA results similar to the statement about cats (for instance, the DNA also unequivocally shows that you are not even close to being a TULIP, or a CHILEAN SEABASS, and so on). However, this will ONLY be made available when you upgrade to our Platinu-mRNA Member plan.

8. Your genetics shows a PLUS (OR MINUS) 15% change in the likelihood of SOMETHING happening to YOU in the FUTURE that is probably very IMPORTANT. Note that there is a correlating PLUS (OR MINUS) 15% likelihood of people around you NOT ACTUALLY CARING about this. Please note the mention of MATH, which in our view further strengthens the validity of this test result.

9. There is a REASONABLY GOOD chance that you are, in fact, a DOUCHEBAG.

Thank you for trusting in our services. If you have any questions or would like to schedule an appointment to discuss your test results in detail, please call 1-800-555-GENE and ask to talk to someone that is wearing a white lab coat.

Sincerely,
GENETIC TESTS FOR REALZ

---

[20] Easily screwed by fake genetic tests.

HOW TO TURN A PR NIGHTMARE INTO A DREAM

Alright everyone, it's time for some major spin control. We managed to plug that pipeline up, but now we seem to be losing the public relations fight what with the freaking amount of bitumen that spilled out. Seriously, the bad press is everywhere, and we are, quite frankly, getting crucified out there. So what can we do about this? How can we turn this PR nightmare into a PR fairytale?

Well, we think that we've got an idea that can't lose. Let me explain. Basically, when we thought about the idea of a PR fairytale, we thought about castles. And when we thought about castles--stick with me here--as vanguards of the capitalist world, of course we didn't think about real historic castles. No, we thought about pink stucco ones, like the kind you might associate with movie studios and animated versions of Cinderella. And then (like magic, we did this all at once, I swear) we said to ourselves, "THEME PARK!" And then we wondered, how much energy is in this leaked tar sand product anyway?

Well, it turns out (with some very speedy back of the envelope calculations) that the amount of energy we can get from it might be good enough to explore the running of our own magic kingdom! Well, at least if we can count on a few more leaks along the way. But how cool would that be? Anyway, here's the gist. We just pull that energy from our happy accident(s), redirect it, and then run this baby! It'll be like the leaks happened **on purpose!** Awesome!

But we digress. Let's not bore you with talk of energy and leaks, let's talk THEME PARK!

Now this is just preliminary brainstorming, but we're thinking a great name would be "Slick City!" Nice, right? Maybe even add to that a catchy tagline - something like **The Family Friendly Pipeline Spill!** We can even have animal characters wandering around the park, with maybe some kind of funky gel-like oil in their fur and feathers so it looks all cool and shiny. And yes, there will be a **Fossil Fuel Palace**, made out of glistening anthracite! I can even envision a theatre area where an oiled-down animal mascot version of the musical **Grease** is performed.

Is it just me, or are people going to pay some serious coin to see that?

And the rides: how about one called “Shutting down the science!” You'd have these carts that go around a track, and the riders have these light guns that shoot at things for points. For instance, they can shoot at all the nasty scientists who want to report on their work, or shoot at research centers that might be making inconvenient discoveries. Ha ha, just kidding - I’m just throwing ideas out there, but you get the picture right?

We also need a giant slide of some sort. What if we design the slide so that it followed the same curve as the hockey stick graph? And what if we call it the Carbonator or something cool like that?

And the big ticket item? Obviously, this will be an epic rollercoaster. Perhaps one made to look like a big old pipeline. We could even make it from real pipeline parts! Don’t we get discounts for those kinds of things? As well, this ride is going to be amazing: it’s going to be the future of log rides. Instead of logs, the folks could sit in oil barrels, and instead of traveling through water, maybe those barrels would even go faster in a petroleum based fluid. Extra bonus if we get to light it on fire!

This is totally a goldmine of an opportunity. It’s like the ideas are just flowing and the theme park is creating itself! FRIED FOOD! Whoa. That one came out of nowhere! Seriously folks, we’ve hit oil here and it’s a gusher.

WHEN CELEBRITIES[21], WHO HAVE BEEN CLONED IN THE MOVIES, GET TOGETHER FOR A MOVIE

SETTING: A Starbucks in L.A. – three celebrities are sitting at a table with their coffees and sharing a newspaper, a fourth is walking towards the table with his coffee.

FADE IN

MICHAEL KEATON
(Approaching the table)
Man, I really need this now.

(Sits down, whilst the others nod or wave).
Is there a free section of paper?

ARNOLD SCHWARZENEGGER
You vant the sports section?

MICHAEL KEATON
Sure.
(Takes the paper and starts looking at the front page)

(A few minutes of silence as everyone reads their newspaper)

HITLER
(Slams his paper down and stands up).
Dis ist terrible! As Fuehrer of the German people and Chancellor of the Reich, I cannot agree with dis. Vee must fight. Neither force of arms nor lapse of time vill conquer Germany. It ist infantile to hope for the disintegration of our people. Mr. Bush may be convinced that America vill

---

[21] Michael Keaton was in "Multiplicity," Arnold Schwarzenegger was in "The Sixth Day," Hitler was in "The Boys of Brazil," and Pikachu was in "Pokemon, the First Movie: Mew vs MewTwo"

win. I do not doubt for a single moment that Germany vill be victorious. Destiny vill decide who is right. One thing only ist certain. In the course of world history, there have never been two victors, but very often only losers.

MICHAEL KEATON
Whoa easy there Adolf. Is that de-caf you're drinking there buddy?

PIKACHU:
Pikachu! Pikachu!

MICHAEL KEATON:
Hey, look at this, Governor Arnold. Looks like you're in the paper today.

ARNOLD SCHWARZENEGGER
I know, isn't it swell? My biceps looked especially good on that day.

MICHAEL KEATON
(sipping his coffee)
Yeah, pretty cool, pretty cool. I've got to ask though, what's it like being a Governor of California anyway?

ARNOLD SCHWARZENEGGER
Oh, you know, nothing special really. Besides, what makes you so sure that it is me and not my clone.
(Everyone chuckles).

HITLER
Hey, I saw Spider Man 2 yesterday – it vas really good. Hey Michael, do dat funny thing I like.

MICHAEL KEATON
You mean this.
(Grabs Arnold by the shirt lapels and pulls him close to his face)

I'm Batman!

ARNOLD SCHWARZENEGGER
(Laughing)
Ya, that kills me too.

PIKACHU

Weeeeeeeeeeeeeeeeeee!

(Darth Vader, the Lord of Sith then approaches the table)

DARTH VADER

Hello Arnold, may I join you?

HITLER

(Standing up and cutting in)

I'm sorry Mr Vader, but dis table ist reserved only for celebrities who have been cloned in zee movies.

DARTH VADER

(Facing Arnold)

Your destiny lies with me Schwarzenegger. Obi-Wan knew this to be true.

(Turning to Hitler, with two raised fingers and speaking very deliberately)

I am welcome to join you for coffee.

HITLER

(In a sort of trance)

You are velcome to join us for coffee.

DARTH VADER

Here, please have my seat.

HITLER

Here, please have my seat.

ARNOLD SCHWARZENEGGER

Darth! Stop that now!

(Hitler shakes his head)

The 'cyborg' coffee group doesn't meet until tomorrow morning.

DARTH VADER

(Turning to Arnold)

Impressive. Most impressive. Obi-Wan has taught you well. You have controlled your fear. Now, release your anger. Only your hatred can destroy me.

MICHAEL KEATON
(Tapping his finger on Darth Vader's arm)
Umm, buddy, I think Arnold told you to get lost.

DARTH VADER
(Looking at everyone)
Hmmmpph, very well.

(Turns away and leaves)

MICHAEL KEATON
(Quietly)
Loser.

PIKACHU
Pfffsssstttt!

FADE OUT

ANYTIME

In some respects, a person reacting to the words "she's in a coma" necessitates a cautioned approach. Is he kidding or is he serious? It's akin to that feeling of discomfort you get when you're not entirely sure if a woman is pregnant or just large. But to me, the phrase represents more than this confusion – it represents a bookmark in my young family's life; despite everything else, it marks an occasion where I think we all grew up.

A week before my daughter turned one, my mother-in-law was a victim of a serious car accident. The physics were horrendous: a large truck bearing down, neglecting the stop sign, and momentum transferred T-bone style onto her small sedan, the lack of side airbags a costly afterthought. In that instant, my wife's mother suffered an injury that would pit her will against a crushed physiology.

And there really is no confusion with the words "traumatic brain injury." Less still when one talks of bruising, swelling or tearing, all possible when an organ like your brain smacks hard against the wall of your cranium. It was also ironic that confusion, a common side effect of head injury, was really the least of everyone's worries. Instead, we were intent on her survival, fearful of the worst, and grateful for the institution that is our medical community. We never saw a light, but we're pretty sure she did.

Meriel was in a coma for three scary weeks, and furthermore remained in the hospital for an additional two months. To be honest, this is a period of time that I recollect with a strange and fitting sense of numbness.

I do remember a few things, though. Of course, the most important memory being that she did survive – a brilliant moment at the time. I also remember my daughter's first birthday, which understandably was an outing of mute proportions – more so, when you consider that our child was oblivious to everything except the brave colours of some wrapping paper. But more than anything (and perhaps this illustrates the whole surrealness of it all), I remember all the driving – driving weekly from our house to my in-law's home, four and a half hours each way; driving daily to the hospital, usually in multiples, back and forth, back and forth. I can tell you that the music mix, trapped in the car during this period, certainly got a lot of play.

Which was o.k., because at least it was a good mix. It was lucky that I go to a lot of effort with my mixes. In particular, the song Anytime by Neil Finn stood out. In fact, it still stands out, and in truth, most anything by Neil Finn does for me. It is as if his well tuned sense of melody is in sync with my own neuronal firings – that is, if we still want to talk about brains and the like.

In this case, however, the song also haunted me. Anytime. A clearer sentiment I couldn't have imagined, especially in the aftermath of what had just happened. It was a sentiment so easy to dwell on, big enough to make you cry, and something that you tried desperately to not connect to your children. It was as if the song perfectly packaged a nugget of wisdom, a much needed warning if you will. And it was telling me in its own beautiful way that we are not invincible. No one is. Not even survivors.

Now, my daughter is almost four, we drive a mini-van, and Meriel is more or less still recovering. It's odd, and I'm not sure how best to describe this, but it's as if she is both present and absent. She is also a constant reminder of what all the doctors forewarned, a formidable challenge on how to deal with lesser expectations. And yes, it is a slow recovery, frustrating even, the kind where sadness chooses to sit and wait with you.

Perhaps this is why the song still rings true for me. It's not as if it's over – survival is relative after all. We are still constantly being reminded by the cruelty of that moment of happenstance, which we now see with double potency in both our past memories and our present eyes.

And yet, despite this, sometimes I think there is still a small fire in Meriel's eyes. Despite the chronic pain, the fatigue, the grayness, I sense that there is, somewhere, the will to fight. Or at least I certainly hope so.

In truth, we could all do with a bit more of that fire, a bit more of that will, no matter how painful or frightening, to make us want to live and carry on. It's like the song says:

"I could go anytime
There's nothing safe about this life
I could go anytime"

No better excuse, really.

THE N.I.H. REJECTS DR. PHIL

Dear Dr. Phil,

Thank you for submitting your application for the director's position at the National Institutes of Health. As the N.I.H. is the principal force guiding America's efforts in medical research, we have strived to consider every candidate's application seriously.

Our first impression was not a good one. You have a loud and exuberant manner that is an oddity in our network of colleagues, and for the duration of the interview process, you were physically sitting on top of our Director (a man considerably smaller than you), oblivious to his muffled and strained murmurs beneath you. We found this quite distracting and wonder what this reflects of your character. Furthermore, although he has only a minor role in the selection process, our Director was quite put out. As the conversation continued, we found other characteristics that troubled us.

Your commitment to, as you call it, "big ideas," whilst commendable, seemed a tad impetuous. Your mention of using your television program or perhaps "your good friend" Oprah's television program to (in your own words) "GIVE FREE GENE THERAPY TO EACH AND EVERY MEMBER OF THE AUDIENCE!" is frankly very unsettling to us.

In truth, we fear that your celebrity status may ultimately impede our principal mandate of excellence in health research. Although some of our members thought it wonderful that you have a Muppet in your likeness on "Sesame Street," your list of other references (e.g., "I drink scotch with celebrities on a regular basis") hardly elicits confidence. To be blunt, your scientific C.V. is poor and your repeated attempts to demonstrate your scientific prowess were laughable at best. (Adjusting the pH in your hot tub does not count, nor does your vasectomy.)

Finally, we found your tendency to talk in meaningless, corny phrases very irritating. Responses like "Sometimes you just got to give yourself what you wish someone else would give you" or "You're only lonely if you're not there for you" are very confusing, to say the least. In fact, our members felt that overall you were even more irritating than the applicant who used the word "testicular" 67 times in his interview.

One member of our hiring committee actually wrote the comment "Who the [expletive] is this guy—Foghorn Leghorn doing Yoda?"

Consequently, the hiring committee regrets to inform you that your application has not been shortlisted for further consideration at this time.

Please tell Ms. Winfrey to stop bothering us.

Yours sincerely,
Dr. Paul Batley Johnson
Hiring Committee
National Institutes of Health

AN INTELLIGENT DESIGNER ON THE COW

Today, I feel like doing a plant – no, an animal. Yes, today, I am going to make an animal. And it will be a masterpiece. I shall call it the.... No wait! Maybe I should think of the name later. Yes, you should always name your pieces after you have completed them. Better that way.

OK then. An animal it is. More specifically, a vertebrate. Large body, four legs, one tail, one head, usual stuff on the head – i.e., let's just follow the standard animalia rubric. Nothing exciting there. Not yet anyway. So let's give it an armored tail, with poisonous tendrils and a stink that can kill. Oooh, I like that – but maybe it's too much. Why such a fancy tail? Maybe the tendrils can come out of its nostrils (note to self: Have I designed nostrils yet?). And the stink can come from the body itself.

But it doesn't quite feel right. Feels forced. No matter, I suppose I can simply start over. Besides, I did the poisonous tendrils last week. But keep the stink? Yes, let's keep that.

I know. How about we give it three, no eleven, no four stomachs! Four stomachs! For the efficient eating, of the grass. I am truly inspired! Don't stop there. How's this? This animal should urinate milk. From its groin, no less. From little appendages which I will humbly call teats that collectively, communally, reside on a mound of tissue I will call a brother.

Now I am on a roll. Milk will flow from the teats of this animal's brother. No wait, I cannot call it a brother. This animal has no lips – don't want it to have lips – too common a thing for a masterpiece. Seen that, done that, yesterday's news. But you can't say the word "brother" without lips. Poor animal, that would be cruel. Instead, let's call it an udder. Yes, an udder – that's much better.

Now, of course, I need to work in a clown somehow. I love clowns. In truth, clowns are my all-time favorite design. How will I do this? Perhaps give the animal a raucous and overt sense of humor? Make it wear funny shoes? Make it scare the shit out of young children? No, not subtle enough – I want this animal to be so much deeper than that.

What if, and I'm just saying things as they come to me, this animal-can-be-ground-and-shaped-into-a-meat-patty- which-can-be-mass-produced-and-fried-on-heating-elements, and-then-sold-by-a-corporate-entity-bent-on-feeding-the-obesity-line-to-young-children -by-using-as-their-public-representation-and-symbol, a-clown, whom-we-shall-call-Jesus (no-wait,-let's-save-that-one-for-later), whom-we-shall-call-Ronald-McDonald, and-these-meat-patties, which-will-be-inexplicably-and-mysteriously-called-hamburgers -after-a-completely-different-animal-I-haven't-created-yet, will-also-be-considered-sacrilegious-by-fully-one-sixth-of-the-world's-population, and-oh-oh-why-is-it-that-the-numbers-O157-cry-out-to-me? because-OH-MY-GOODNESS-I-can't-believe-it, but-this-stuff-is-just-so-brilliant!

…

Take a breath. WHheeeew-hooooooo. Calm down. That's pretty good. But maybe just think about some of the simple things now. Like color. Yes, color is good. And easy – let's go with the rustic look, plus spots. Et voilà. We have finished yet another creation, which for some reason, I feel inclined to call a cow. Hold on, one last thing. It shall go "moo" when it speaks. Yes, that's a nice touch, even if I do say so myself. People are sure to talk about that one, maybe even create a song or two.

SIMILARITIES I NOTICED BETWEEN GEORGE W. BUSH AND THE BURNING BUSH

Natural disasters figure prominently for both.

In their own ways – doing their part to increase carbon dioxide levels in the atmosphere.

Neither necessarily follows international conventions

I gather, both on the same page with this stem cell business.

When they speak, it's kind of surreal.

ON SCIENCE, BEER, AND THE NUANCES OF PROGRESS

It was the summer of 2008, and the four of us, all scientists, were admiring the fare on the table. We were sitting just outside of a hot dusty shack that was deep in the lanes of Ibadan, one of the largest cities in Nigeria. And although we were waiting for what was promised to be "the finest bowl of fish pepper soup on the planet," at that particular moment, we were actually more focused on our beverages.

Dr. Oyekanmi Nashiri (or Nash, as he preferred to be called) leaned over and placed a coaster on top of my glass, "The coasters are for the flies," he explained. " Nigerian flies like beer." Nash was our guide, and a local biochemist bent on improving science in his beloved Nigeria, restoring it to its glory days before military rule and corruptive practices had whittled its infrastructure to its current challenged and primitive state. He was convinced that drinking beer with foreign scientists like myself was part of this process.

We were drinking Gulder, an iconic Nigerian lager omnipresent in the glasses of the locals as well as many a billboard extolling its virtues as "The beer for the real man." Both my colleagues, Michelle Brazas, a bioinformatician, and David Peterson, a specialist in *Plasmodium* biology, thought this tag line amusing. Michelle, in particular, wondered if there was a "beer for the real woman." The Gulder was good, flies notwithstanding – the taste was crisp and light, very appropriate for the hot and muggy weather – perfect even for our spicy main course (which was delicious, as advertised).

The past few days, we had conducted an unscientific survey of Nigerian beer, with Gulder being the latest specimen added to our collection. The others included Star Lager, Legend Extra Stout, and Grand Lager, as well as bottles of Guinness and Guinness Extra Smooth. Nash also mentioned that there were several other brands that we hadn't yet sampled, such as Mopa and Hi Malt. The selection was impressive, and I was curious, "Where did all this variety come from?" I asked Nash.

"Right here, David!" Nash extended his arms outwards, "We make many a fine beer here. You know, I believe we are even home to the largest Guinness brewery in the world!"

Later on, I found out that Guinness does, indeed, have a huge brewery here. As well, the Gulder we were enjoying was produced by Nigerian Breweries Public Limited Company (or PLC), a subsidiary of Heineken. In fact, the business of beer in Nigeria was very good. At that time, Nigerian Breweries PLC and Guinness Nigeria ranked second and sixth in net worth out of all West African businesses.

This seemed ironic given the place where we were. The usual perspective is the one that is illustrated with worrying statistics and figures – like the one that was ingrained in my own consciousness where 2006 figures had suggested that 71% of the Nigerian populace was still subsisting on less than $1 per day. And whilst hallmarks of such poverty were certainly evident around us, signs such as our choice of beers, the ubiquity of cell phones, and the slow but noticeable improvement of science facilities we had seen over the years suggested a different story. "Things are changing," Nash would say, "There is a growing middle class, and hopefully, good things are to come." And then he would add, "Did you know the President Umaru Yar'Adua was once a scientist, a chemist?"

Today, it's 2014, and the statistics on Nigerian standards of living have improved, albeit slowly. Policy makers can now proclaim a welcomed downward trend, where 61% of the populace are now living on less than $1 per day. Still, Nigeria is fraught with challenged optics: most notably around physics education, kidnapped girls, and the frightening actions of Boko Haram - the name of which when translated from Hausa says it all, "Western education is sin." It would seem that Nigerian science still has many obstacles in its path. Except that I am made hopeful by a memory from that fish pepper soup meal.

I remember looking down at my Gulder bottle, and seeing that the label clearly informed me that it had an "alcohol by volume" content of 5.2%. And implicit with that was the notion that, yes indeed, somewhere in this land was a fermenter – apparently many and some very big fermenters. Beer making, after all, is very much an application of science, and in academic circles the act of brewing is most often cited as the earliest form of biotechnology. I didn't know it at the time, but the two or so litres of Gulder we consumed that evening most likely came from only 400km away, part of the 300 million litres of beer produced annually by the Ama brewery in Enugu, one of the largest and most modern facilities in the world.

This was a marvelous moment – a clear realization that when all things are considered, the world is strangely nuanced. Here, at that moment, a fermenter was everything opposite to what I saw around me, and in hindsight, also opposite to what I read in the news today. Fermenters need to be accurate, precise, sterile, proficient, and safe - they are clean, big, shiny, and in their own way, elegant. In that moment, beer represented a microcosm of the very real progress of places like Nigeria – the same progress that will hopefully help Nash realize his dream of restoring science to his country. That Gulder in my hand was strangely hopeful.

I remember at that moment, Nash nudged me from my thoughts and asked if I wanted another beer. I nodded. "Do you feel like something stronger?" he asked. "If so, it is time we try the Gulder Max."

AN OPEN LETTER TO THE HUMAN RESOURCES DEPARTMENT OF THE SUPERFRIENDS

Dear Mr. Superman, Mr. Batman, Ms. Wonder Woman, and other esteemed do-gooders,

Although I have been waiting patiently for a few years in the hope that an advertisement would appear, I feel for the sake of my career that now is a good time to approach you. In essence, I am wondering whether you are, or will be, accepting any new members into your fine organization. More specifically, I am inquiring as to whether you need the services of a geneticist, since that is my particular field of expertise.

Part of the reason I am contacting you now is that I suspect you are possibly hurt by the unwarranted waning in public interest associated with your group, a symptom that likely correlates to the soaring popularity of some of your competitors—the X-Men and Spider-Man, to name two.

Anyway, this is why I think I can help—as a geneticist, I can bring a lot to the table. To me, it's no coincidence that the X-Men, Spider-Man, and the like are mostly a consequence of fortuitous genetic modification.

More specifically, my knowledge in genetics can directly tackle your weaknesses. For instance, current genetic technologies could be utilized to offset Mr. Superman's annoying kryptonite problem. Whether it's the result of something specific in *his*genetic makeup or the result of the rest of society having some sort of innate immunity, the issue at stake is a difference in biological makeup. This, of course, makes it a perfect candidate for targeted gene therapy.

Mr. Batman could also benefit greatly from a genetic analysis. I would not be surprised if his manic-depressive tendencies are hard to manage and counterproductive to the group as a whole. Here we can perform a few diagnostic genetic tests, which can then go to great lengths to effectively pinpoint and manage these potentially bothersome tendencies.

Even Ms. Wonder Woman could stand to gain from my genetic expertise. Clearly, engaging in intensive hand-to-hand combat with her sizable chest is problematic. But how exactly would you perform breast-reduction surgery—or any surgery, for that matter—when an individual's magic bracelets are constantly maneuvering to defend against an incoming scalpel? My point is that you don't have to—I may be able to do something about this by using current genetic-manipulation methodologies.

And just imagine what could be done with cloning. The mind reels, does it not?

Now, with respect to fighting crime, I think that, overall, it wouldn't take much effort to transform me into a fully functional Superfriend. I already have a well-equipped laboratory facility, which, with your help, could easily be relocated to the appropriate underground cave, glacier, secret island, or space station.

As for a costume, I own several lab coats, which, when worn with a good pair of spandex pants, will, I'm sure, sufficiently instill some semblance of fear into those who choose to do battle with me. I would offer to wear some retro-looking goggles as well, but, unfortunately, I need my prescription glasses, and, well, contacts tend to make my eyes itch.

Although I can't fly, and I don't own anything close to resembling a Batmobile or an invisible jet, I do drive around in one of those stylish yet practical Mazda MPVs. If you recall, this is Mazda's very popular minivan model (you know—zoom zoom!), which would probably look quite superheroish were I to paint some lightning bolts or DNA strands on its sides.

Also, if it helps, I know quite a few physicists who may be able to lend a hand with Ms. Wonder Woman's embarrassing "the jet is invisible but I'm clearly not" situation.

Yours sincerely,
Dr. David Ng
University of British Columbia
Vancouver, BC, Canada

A UNIVERSITY JOB POSTING (OR BECOMING A PROFESSOR IS HARD THESE DAYS)

This is a call for outstanding candidates to apply for a tenure track assistant professor position within the context of the Department of [*subject name*] at the [*institution name*]. The successful applicant is expected to work in areas of interest to current faculty members, to interact with related groups within our network and to have demonstrated ability in producing research material of excellent quality and interest.

Due to the competitive nature of this process, we ask that all candidates at the very least meet the following criteria:

The candidate's current area of specialty must contain at least fourteen syllables.

The candidate's expertise must speak naturally to collaborations with the disciplines of science history, Jungian philosophy, international peacekeeping, French Canadian politics, molecular genetics, early 80s pop music criticism, and West African cuisine.

The candidate must be able to "flex arm hang" for a minimum of twelve minutes.

The candidate must exhibit no more than two degrees of separation from Kevin Bacon.

The candidate must be able to rub their tummy and pat their head at the same time.

The candidate must, in no uncertain terms, smell nice.

In addition, short listed candidates will be subjected to a rigorous interview process that will likely involve puppetry, ultimate fighting, and some interpretative dance techniques. This, of course, might be televised nationally on television, so it is advisable that all applicants prepare in advance for these skill sets.

The successful applicant will covet a salary that will commensurate with experience and research record, but realistically is dependent on an obligation to play as the principle string in the University's Chinese Orchestra during the first three years of his/her track.

We will also endeavor to provide the applicant with reasonable research space, and note that we have one of the country's best supply of camping gear, should this be an issue. We do however ask that successful candidates will themselves provide start up funds to the sum of $1000, which must be used within 48 hours. During that period, you will, of course, be wearing brightly covered overalls and have access to a skilled carpenter who will almost certainly be just as attractive as you.

The [*institution name*] is one of the leaders in North America with strong connections with many well regarded institutes, and we look forward to continuing this tradition with this placement. We hire on the basis of merit and are committed to employment equity. We encourage all qualified persons to apply; however citizens and permanent residents will be given priority. No losers please.

EUPHEMISMS THAT ALSO SOUND LIKE STRANGE TISSUE ENGINEERING PROJECTS

Banjo On My Knee

Bleeding Heart

Foot In Mouth

Dick Head

Shit for Brains

Get this Monkey off my Back

He's a Leg Man

Space Between The Ears

Baby Snacks

White Meat Only

Biggest Asshole

IF ORBITALS WERE TO PLAY ACIDS IN A GAME OF BASKETBALL

I wrote this years ago, when my friend Ben Cohen and I, cohosted a SCIENCE SHOWDOWN during the 2007 NCAA. This was where different scientific terms were pitched against each other and it was seriously geeky. Still, we had many folks play along, writing some really excellent creative science pieces – you should check them out as they're lost on the internet somewhere.

* * *

Welcome folks, to this here what we'll call the beautiful game (at least we'll say that for the molecular level). This game really had it all, it was dynamic, it had equilibrium, it had fluid transition, and it was catalytic. It involved freakishly large chemical sounding words, and also a weird scoreboard that looked something like this:

	s	*p*	*d*	*f*	*g*
1	1				
2	2	3			
3	4	5	7		
4	6	8	10	13	
5	9	11	14	17	21
6	12	15	18	22	26
7	16	19	23	27	32
8	20	24	28	33	38

But hey, whatever, right?

The game started off slowly enough, with Team Acid moving the ball well. Their game plan was fairly straight forward, and with a play by play that looked a little bit like this:

$$HA(aq) + H_2O(l) \rightleftharpoons H_3O^+(aq) + A^-(aq) \qquad K_a$$

But then the d-orbitals stepped it up in strides. Moving with both precision and with uncanny diffuse footwork that was seemingly hard to defend. Full of spark and basically responsible for a lot of the colour of the game, their floor plan followed a few extravagant patterns. Such as:

this,

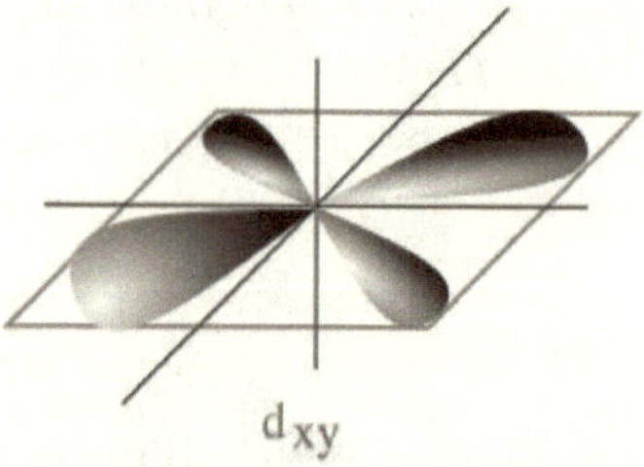

this,

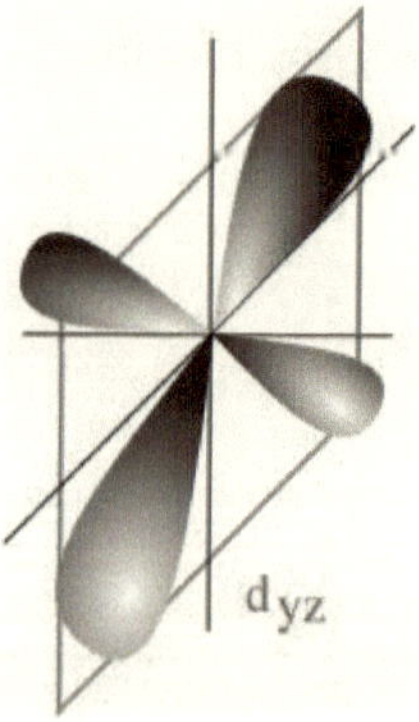

this,

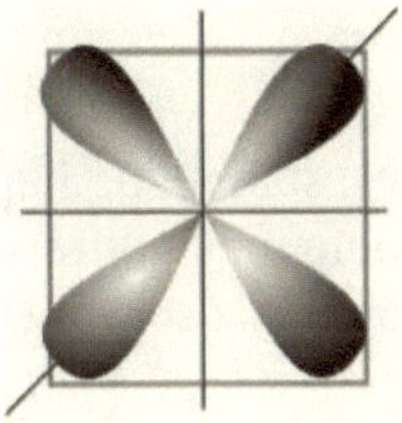

d_{xz}

and every now and then, even this:

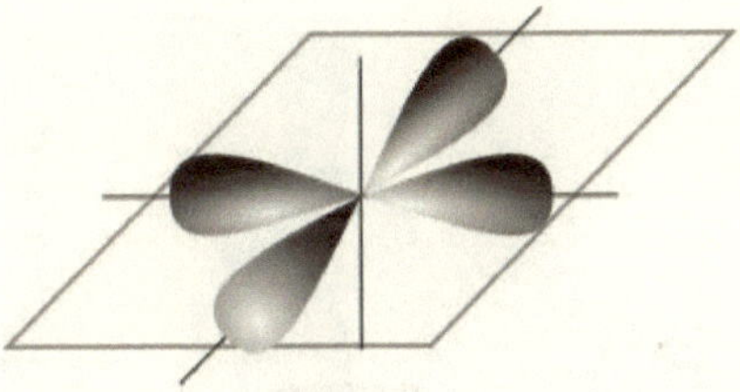

$d_{x^2-y^2}$

But the coaching staff for the Acids were a smart bunch, and soon realized that the d-orbital gameplans were essentially all similar. That for all of their flash, (as well as support from the well resourced transition metals), they were (as the coach was heard to say), “just moving about in the same f**king four way formation – any sh*thead should be able to mess that up!”

So, newly motivated, they countered with:

$$H_2A(aq) + H_2O(l) \rightleftharpoons H_3O^+(aq) + HA^-(aq) \qquad K_{a1}$$

$$HA^-(aq) + H_2O(l) \rightleftharpoons H_3O^+(aq) + A^{2-}(aq) \qquad K_{a2}$$

and on occasion this,

$$H_3A(aq) + H_2O(l) \rightleftharpoons H_3O^+(aq) + H_2A^-(aq) \qquad K_{a1}$$

$$H_2A^-(aq) + H_2O(l) \rightleftharpoons H_3O^+(aq) + HA^{2-}(aq) \qquad K_{a2}$$

$$HA^{2-}(aq) + H_2O(l) \rightleftharpoons H_3O^+(aq) + A^{3-}(aq) \qquad K_{a3}$$

So for a while, it looked like Acids had it in the bag. That they would advanced to the final eight. That it was all over for Team d-orbitals, who would then have to delocalize[22] and do other things that incorporated some kind of sad chemistry pun.

But then, in the last ten minutes, something happened. Something beautiful, something different, but not something entirely unexpected – because let's face it, it's easy enough to google this sort of thing these days.

And so, here's how it happened. This, my friends, is what the game will be remembered for:

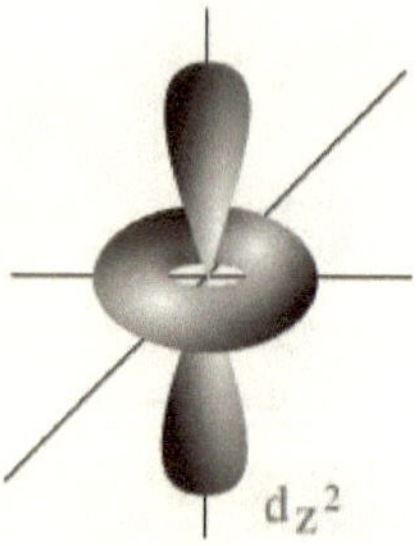

Where really all we can say is, SWEET JESUS!

And then it was basically all over – Acids were all but destroyed from that point on. And this, here, is the final result – d-orbitals takes the game. In style some would say, and by a score of 98 to 86.

* * *

Later that evening, I caught up to the Acids, with some questions. But they had no answers, just a few sorrow shrugs and some parting words.

$$HCl(aq) + NaOH(aq) \rightarrow H_2O(l) + NaCl(aq)$$

FIN

[22] See http://en.wikipedia.org/wiki/Transition_metal

GRIMACE SPEAKS TO A GENETICIST

GRIMACE: What am I?

GENETICIST: That is a very interesting question indeed. And we should begin by briefly discussing your known history. According to your records, you were born as "Evil Grimace," with four deft arms, and a penchant for amusing yourself by stealing milkshakes from small children. Then, in 1974, you experienced a change of heart, a loss of two arms, and a metamorphosis into what is your current incarnation—a supposedly warm, gentle, and seemingly living representation of the "embodiment of childhood."

GRIMACE: Is that why I have only one orifice?

GENETICIST: Perhaps so, as childhood is a period marked by the most basic of bodily functions. In truth, it is that kind of interesting nuance that makes me suspect your being a genetically modified organism. Furthermore, the timing of your appearance coincides perfectly with a social phenomenon during the '70s. A time when discussing human cloning was culturally fashionable, when books like *The Boys From Brazil* and *In His Image*
appeared on bestseller lists.

Also, you are purple like a giant areola.

GRIMACE: How can I find out more?

GENETICIST: A promising course of action is to try genetic counseling. Which, in the conventional sense, suggests that we investigate your network, both in family and in friendship. This is to help construct a more complete picture of your being and, more importantly, your past. From this, we will have a firm starting point from which to build.

GRIMACE: But I have no family, no real friends, and Ronald, frankly, scares me. What other alternatives do I have?

GENETICIST: Ronald scares us, too, but that is for another interview. Under those restrictive circumstances, one possible alternative is to contact non acquaintances with similar traits. Perhaps someone like Barney the Dinosaur, who is also big, purple, and waves a lot like an idiot. Similarly, we could simply forge ahead and arrange for a genetic test. This is a process that will allow us to peer at your very own genetic code, and is something that will surely resolve the mystery that surrounds you.

GRIMACE: Like why I am so popular with the ladies?

GENETICIST: Yes, exactly! In some respects, you could be the perfect metaphor for what is both wonderfully right and terribly wrong about genetic manipulation. Due to the marvels of this technology, you appear to have luxury, wealth, fame, as many women as you desire, and yet you have no identity, no origin. If ever there were such a thing, you are an organic black box.

GRIMACE: I think it's because the ladies like my massive tongue.

GENETICIST: Which is magnificent indeed! In fact, seeing it now, I am struck by how similar your appearance is to that of a tongue, *a taste bud*, to be specific. To entertain this avenue, I ask that you take a moment to study and answer these five carefully designed questions:

(1) Do you find that you sweat profusely such that you are always, to a certain degree, moist?

(2) Do you find yourself a constant victim of paper cuts, specifically when handling your letters of correspondence?

(3) Do you find you enjoy bathing in scented waters but are repelled by thoughts of swimming in the sea, perhaps fearing that the salt will further constrict your already-tender skin?

(4) Do you notice that when you are jumping on a trampoline, the consonant sounds "l," "n," "d," and "t" appear as if by magic?

And (5) Do you, during the winter season, always find yourself inexplicably and inconveniently stuck to cold metal structures?

GRIMACE: Hmmm, maybe the trampoline one, but otherwise, no.

GENETICIST: Ah, well, it was only a hypothesis. It appears that we will order that genetic test after all. But first, I feel compelled to present this stern warning: these tests can be excruciatingly accurate sometimes. You may, quite frankly, be disappointed with the result. You see, I cannot control the outcome of the test. I do not possess that power. I am not God. I am, sadly, only a geneticist.

EXCERPT FROM "SHORT PROTOCOLS IN MOLECULAR BIOLOGY: MAD SCIENTIST EDITION"

1. Vortex each overnight bacterial culture thoroughly, and transfer 1ml into a clean microcentrifuge tube. *HA HA HA! SOON, MY DARLING BACTERIA SOON!!*

2. Spin the cells down for 30 seconds at maximum speed in the microcentrifuge. Remove all of the supernatent by pipeting out the last bit of media left. *YES... YES... YES!!*

3. Add 200ul of STET buffer to your cells and resuspend by vortexing briefly. Then add 20ul of 20mg/ml lysozyme solution in 25mM Tris-HCl, pH8. *THEY MAY HAVE MOCKED ME, BUT NOW THEY WILL BE SORRY, SO VERY VERY SORRY!! HA HA HA!!*

4. Vortex briefly again, and place tubes in a boiling waterbath for 60 seconds. *SOON THE POWER WILL BE MINE!! ALL MINE!!*

5. Immediately spin for 5 minutes at maximum rpms in the microcentrifuge. *HA HA HA!! THEY WILL PAY!! THEY WILL PAY!!*

6. Remove the pellets (chromosomal DNA) with the flat end of a sterile toothpick. You will find that the pellet will stick nicely to the toothpick without mixing in with the supernatent. Discard the pellet. *THE FOOLS! THE FOOLS!!*

7. Add an equal volume of ice-cold isopropanol to the supernatent. Mix. Chill for 10 minutes at -20°C. *THE IDIOTS SAID IT COULDN'T BE DONE, BUT NOW WE'LL SHOW THEM!!*

8. Spin down DNA precipitates for 5 minutes at maximum speed in the microcentrifuge. Wash pellet with 250ul of 70% ethanol. Spin down for 1 minute. Remove the supernatent. *THEY'LL BE SORRY THEY EVER DOUBTED ME!!*

9. Air-dry the pellets or dry in SpeedVac for 10 minutes. Dissolve the pellet in distilled water or a Tris EDTA buffer. This is your plasmid DNA preparation. *HA HA HA!! NOTHING CAN STOP ME NOW. MUWAHAHAHAHA! MUWAHAHAHAHA! ...*

BE VERY AFRAID

(This is from a speech in 2006)

A few months before he died, a Nobel Prize winner wandered into my office, sat down, and proceeded to talk about science and ethics. He did this for about an hour. In fact, most of it boiled down to something like this.

"Science is in a very interesting predicament these days. It has accelerated so much, in so little time, and has led to a glut of information. It has progressed beyond our wildest dreams, such that we can do amazing things, exciting things, even frightening things."

Of course, he said this stuff in a much less verbose way. The Nobel Prize winner was Michael Smith, inventor of something known as 'DNA site-directed mutagenesis,' an unfortunate mouthful that some view as the first step towards all this genetic modification brouhaha. You should know, however, that he was a very nice man.

Nevertheless, it got me thinking about fear and science in our world, and six years later, this is what I've come up with.

We should be very very afraid. That, incidentally, is one word for each year. I'm using a proverbial *we*, of course, since I'm of the opinion that it's not necessarily the science that is faltering. There is nothing to fear from the science itself. Remember, science and understanding is progressing right? This proverbial *we* really means all of us in general.

* * *

Here's something interesting I bet you didn't know. My Dad beat up Bruce Lee. This, I swear, is the God's honest truth. Mind you, he was ten at the time, and Bruce (I believe) was maybe seven or eight. Anyway, what I just did, my friends, was what PR folks would call a *spin*. Now spin is something we should fear.

Spin factors in on stories like GM foods, a big part of that brouhaha mentioned earlier. Did you know that the pollen from certain GM crops has an inadvertent and toxic effect on Monarch butterflies? You probably knew this already since it was pretty hard to not notice WTO protestors dressed in tatty butterfly costumes on television.

More specifically, all of this fuss was over a GM variety known as Bt corn. Which, by the way, is not short for Bag O' Treats, or Beelzebub Tamer, or the Big Tickle, or even Bitchin' Toxin for that matter. In truth, it is short for *Bacillus thuringiensis*, a small and relatively unassuming bacteria.

Here's why.

Bt corn has had a special gene inserted into its make-up.
This, incidentally, is a gene from the Bacillus.
Which, incidentally, is responsible for producing something known as a crystal delta-endotoxin.
Which, incidentally, is often shortened to *cry*.
Which, incidentally, has the power to kill insects that might otherwise make your corn look a little spotty.
Which, incidentally, is the same stuff that is commonly used as pesticide in organic farms.
What a marvelously ironic world we live in.

But back to the butterflies. Did you know that the aforementioned toxic effect was a result of force-feeding the butterflies roughly 5 times the amount of pollen that they would have ingested in the wild? Sort of like drinking 40 cups of water a day.

I think if I was the butterfly, I would be angry and protesting as well. The point is, of course, spin. Force-feeding butterflies, even to make a point, is hardly what you would call environmentally friendly is it?

And what about those big ag-biotech firms espousing their commitment to stem the tide of world hunger? To which I say, give me a break! I have a feeling that deep down they have other priorities in mind.

* * *

Or you could say that they don't care- better still, call them apathetic. Or, like teenagers, maybe they're just acting cool.

Arguably, 52% of new car owners think they're acting cool. This number reflects the millions of SUVs and light trucks that were sold in the US last year. One reason I make this statement is because of an interesting story called "Car Wars." A short version goes like this.

Figure Title: Collision Aggressivity

		You Drive	
		"Civic"	"SUV"
I Drive	"Civic"	Better for both	Your kids ... live
	"SUV"	My kids ...live	Worse for both

A longer version goes like this. You are buying a car. You are not thinking about the damage a bigger car causes. You are not thinking about the sense of security you feel because you are higher up in your bigger car. You are not thinking about the car's subsequent higher centre of gravity. You are not thinking about previous year's 14% increase in rollover fatalities. Instead, you are thinking about the 11 cup holders a 2005 Ford Expedition has. You are thinking how nice it is to have the option of traversing rough terrain, no matter how unlikely. But most of all, you are thinking how the car looks "really kick ass"

But wait, another reason I call these car owners apathetic is because of something called global warming. Bigger cars do, indeed, cause bigger damage. About 85% of the energy used in the States is derived from fossil fuels. 40% alone from the use of oil, most of it from automobiles, and most of that not from Honda Civics. All of that CO_2 from your friends' SUVs, leading to warmer climates, el ninos, rising seas, bad bad weather and the like. Who could have guessed that being 'cool' could be so scary?

* * *

I'm assuming that you didn't know that I have two young children. They mean the world to me, and I, like other parents, try very hard to do the right thing. Like not freak over buying a Tickle Me Elmo, for instance. Have you ever seen a stampede of parents? It is altogether a very surreal and somewhat ugly sight. Sometimes that happens when you decide to not be apathetic and actually want something but it is no longer there. Call it a form of withdrawal if you will.

Our oil is supposed to run out sometime in the next 100 years or so. And this is loosely based on current consumption figures, meaning that this number could be significantly smaller. Which would be a problem, since we seem to be very very fond of oil. It might even start a war or two.

The usual alternatives frankly don't cut it right now.

Fuel cells, akin to batteries with gas nozzles, are disappointing. Most use hydrogen. Lovely in that water is the only pollutant being produced, but cruelly inefficient in that producing the hydrogen in the first place is energy expensive (in fact fossil fuels are generally used to do this). People also think hydrogen is sinister, wrongly so as it were, for being ballistic. I guess people forget that petroleum, itself, can explode.

Hydroelectricity, on the other hand, does not go boom, but needs a river that does not object to being dammed – a feature, which is all the more troublesome in the chassis of a car.

Solar still relies on complicated, expensive, nasty chemicals. Please say it with me – solar still relies on complicated, expensive, nasty chemicals. And what about wind energy? Wind powered turbines kill innocent birds – do I really need to say more?

But wait, there is always banana power. Yes, those crazy Australians have figured out a way to use banana waste to generate electricity. Apparently 60kg of decomposing bananas produce enough methane to power a fan heater for 30 hours. Incredible how useful this might be with our screwy weather.

* * *

Here's what Dole has to say about bananas. "Bananas are the most popular fruit in America. An average person eats 29 pounds a year." Nowhere does it tell you that bananas are reproductively sterile. A bit ironic, really, given its overly phallic shape.

Fact of the matter is that the bananas we eat are clones of each other. That's great! No seeds! That's bad! One banana being susceptible to disease means all bananas are susceptible. And because there are no seeds, it's very difficult to breed better varieties.

Here are some biggish words to throw at you. *Mycosphaerella fijensis*. These two refer to a fungus that causes a banana disease known as *Sigatoka*. Believe it not, these words have been useful to me at the occasional cocktail party. I mean really, doesn't even saying them give you the tingles? Of course, if you were a banana, you might actually wet your pants.

Some have even said that the banana could be wiped out in as little as 10 years time. Look out for the fruit lover's stampede!

* * *

So this is what people are doing.

They're inserting genes into bananas to make them stronger. And maybe, one day, they'll insert genes so that you can grow them anywhere, even in colder places like Winnipeg. Imagine that – a prairie banana.

Except that if this was done, I'm sure many farmers in Central America and the Caribbean would be royally pissed off. Worse still, they would go bankrupt. How can they compete with Winnipeg?

This sort of stuff was already happening in 2004, a year which (in case you didn't know) the United Nations had declared "The International Year of Rice." I guess since some big biotech firms sequenced the rice genome in 2002, a lot of patents have been issued, a lot of info passed around, some say inappropriately. No matter, what we now have are American versions of Basmati and Jasmine rice. Pity the poor folks in India and Thailand who have to compete.

Money, of course, makes the world go around. I wish I could take credit for that statement, but I can't. I also can't take credit for the following Oscar Wilde quote:

"It is better to have a permanent income than to be fascinating."

Which, in its own way, says it all perfectly. That is, economics rules our society. To survive, people very quickly learn to look out for the bottom line, and pray that they never ever get caught in the crossfire.

Which is difficult these days since we're shooting all over the damn place. Because of science, the turf we stand on may not be so special anymore, which is frightening if you thought your turf was special for growing bananas, or jasmine rice, or sneakers, or weapons, or anything really.

* * *

Guess what? Earlier on, I sort of told a small lie. Turns out, I didn't even mention nuclear energy as an alternative and I should have. Nuclear energy is clean, abundant, and, most importantly, technically doable. Here's what a colleague had to say about it:

"Generating nuclear power is as cheap as it gets. Unlike coal and natural gas plants, there are zero emissions — no CO_2, no sulphur and nitrogen oxides. No greenhouse gases, no ozone depletion, no acid rain. No rivers to dam up, no valleys destroyed to build your hydroelectric plant."

Of course, he also mentioned that, "when the nuclear fuel is spent, the by-products are highly radioactive – this is very bad."

Which is why in Nevada, specifically at Yucca Mountain, they are digging the mother of all holes. Home sweet home, as it will be, to most of the 77,000 metric tonnes of nuclear waste our neighbours down south have already generated. Just *most*, not all, not even with space left over for future sakes. Price tag? To date, it's cost about US$60 billion dollar, with another US$7 billion to study and/or argue over it. After all this, there is also a distinct possibility that it won't work or won't open.

And one single location! Can you believe it? Imagine lead plated trains and trucks going through your community, on "their way" so to speak. Or maybe it's best that they didn't go through your turf in the first place.

* * *

Of course, Nevada does have its advantages. It has some very specific laws, no doubt carefully thought out for the common good. For instance, it's illegal to drive a camel on the highway, but it is still legal to hang someone for shooting your dog on your property. By the way, human cloning is also legal in Nevada. In fact, when I last checked maybe a year or so ago, it was only banned in 8 U.S. states.

Don't get me started on human cloning. Creating life in this fashion is still a creepy proposition to most of us here at science land. But isn't it wonderful that advocates had Rodney Dangerfield and Raelians as spokes people. What could be more perfect?
And what about therapeutic human cloning? Superman, or Christopher Reeve to be exact, was more or less right – the use of embryonic stem cells for things like treating a plethora of tricky diseases, or regenerating tissues of all types has great potential. But what exactly is an embryo, a stem cell? Where does "life" begin exactly? Instead of talking about the "The Facts of Life" will we instead need to talk about "The Facts of STEM CELL / ZYGOTE / BLASTOCYST / EMBRYO / FETUS / NEWBORN?"

Maybe we should ask a creationist. Actually, better not – that would be stupid.

Instead, maybe we should vote on it. Maybe then we can come to a decision.
Because voting does that sort of thing. It gives you a small say about how things are determined, although sometimes, you may still disagree with the end result. Still, I suppose that's much better than in China. Nobody votes per se in China, so regulation just happens.

In China, couples are strongly advised to give birth to only one child. Most people, I think, already know this. Most people, however, don't know that before a couple becomes a couple, they have to take mandatory genetic tests. This is to see if they are compatible. In a sense, to question whether the child they may produce has a significant risk of developing something bad. Here it is in plain English:

Article 10: Physicians shall, after performing the pre-marital physical check-up, explain and give medical advice to both the male and the female who have been diagnosed with certain genetic disease of a serious nature which is considered to be inappropriate for child-bearing from a medical point of view; the two may be married only if both sides agree to take long-term contraceptive measures or to take ligation operation for sterility. (China's Maternal and Infant Health Care Law, 1995)

By the way, this is called legislated eugenics, which some people think is dangerous. But I ask you, in light of the one baby rule, is it really?

* * *

In the end, I guess you have to vote for a government that you think will make the right decisions. Hope they have the knowledge to make informed choices.

But listen people. Let's be serious. When everyone is busy spinning, acting cool, washing their SUVs, rearing children, eating bananas, gardening in Winnipeg, making money, diverting traffic, watching TV, taking tests, creationing, and voting for government, who has time to stay informed? And that, my friends, is perhaps the scariest thing of all.

- - -

Thanks to Peter Danielson and Stephen McNeil, respectively for discussions on SUV's and alternative energy sources.

NOTES FROM MATTEL'S "FUTURE OF BARBIE®" BRAINSTORMING SESSION

Concept: Hybrid Barbie ®
Description: Barbie doll powered by both conventional gasoline engine, as well as an electric motor.
Pro: Barbie is emissions-compliant.
Con: No one can figure out a good place for the gas nozzle to go in. It always ends up looking inappropriate.
Potential slogan: "This baby gets up to 40 miles per gallon."

Concept: Stem-Cell Barbie ®
Description: Produce a plastic mesh form in the shape of a Barbie doll. Seed this mesh with embryonic stem cells. Culture in bio-chambers until cells infiltrate and coat the plastic form with skin tissue.
Pro: This Barbie might heal itself.
Con: This Barbie might get cancer
Potential slogan: "Feels like real skin because it is real skin."

Concept: Schrödinger's Barbie ®
Description: Interactive Barbie doll placed inside a thick lead box, containing a mock cyanide canister, and mock Geiger counter. The Geiger counter may or may not release one decaying mock atom, which in turn, may or may not break the canister releasing the cyanide. Therefore, child would be uncertain as to the fate of the Barbie doll (who could be pretend-dead or pretend-alive) until the lead box is actually opened.
Pro: This is fun way to illustrate an aspect of quantum law, which suggests that due to the superposition of states, Barbie is both dead and alive until the box is opened.
Con: Huh?
Potential slogan: "Schrödinger's Barbie—be the first to give a shit."

Concept: Super Malleable Barbie ®
Description: Produce Barbie dolls using the Dow Corning 3179 dilatant compound (a mixture containing silicone oil and boric acid, commonly known as Silly Putty).
Pro: Barbie can bounce.

Con: When Barbie pretend-falls asleep whilst pretend-reading a newspaper, the newsprint will show up on her face.
Potential slogan: "Ken will thank you."

Concept: Flame-Retardant Barbie ®
Description: Coat existing doll product with copious amounts of the common flame retardant, polybrominated diphenyl ether.
Pro: Excellent opportunity for accessories (fireworks, matches, flame throwers, etc).
Con: Excellent opportunity for accessories (fireworks, matches, flame throwers, etc).
Potential slogan: "Throw the Barbie on the barbie!"

Concept: Supercomputer Artificial-Intelligence Robot Barbie ®
Description: Multiple clusters of high-powered processors networked to a Barbie doll mainframe. 2 USB ports standard. DVD/Blue Ray player drive optional.
Pro: No more stupid brainstorming sessions—send Barbie instead.
Con: Small chance of total world domination and destruction of the human race as we know it.
Potential slogan: "Kicks ass at chess!"

WAYS CHARLES DARWIN COULD JUMP THE SHARK

He makes exploitative prime-time-television cameos

Darwin appears as the "man in need of a haircut" in an episode of *Gossip Girl*, and is the first contestant voted off *Survivor: Galápagos*. Eventually makes Barbara Walters's list of most fascinating people of the year. Cries during interview.

He sells Darwinian fashion accessories

Darwin takes advantage of the current interest in natural fashion products. Begins marketing items like organic-cotton neckties and honey-flavored lip gloss. Darwin-sanctioned "stylish feces beads" appear soon after.

Reviews movies

Darwin becomes a staff writer at *Rolling Stone*. Is asked to critique animal-related movies. Loses credibility when caught gushing over *Catwoman*.

Hosts free holiday cruises

Moonlighting as an authority on nature and boats, Darwin takes advantage of free holiday cruises. His cruise talks are very successful and Darwin becomes the No. 1 hit when Googling the words "lido deck."

Guest-stars on *The Dog Whisperer*

Memos to the effect of "He sailed on a boat called the *Beagle*. Isn't a beagle a dog?" begin to circulate. Capitalizing on this attention, Darwin joins *The Dog Whisperer* as a dog-anatomy expert. Wins Emmy for segment on comparative dog-tail structure. In acceptance speech, cites the fact that "dog" spelled backward is "god" as evidence of a Christian conspiracy.

Joins the Ice Capades

Darwin is hired for small part in a *Lion King*–themed ice show. Takes skating lessons and practices hard. Soon nails both the triple axel and the triple lutz. Is fired from the show when he tests positive for performance-enhancing drugs.

Loses his mind

Darwin endorses Scientology.

Dresses up as Santa Claus during holidays

Sporting his full white beard, Darwin is hired to impersonate Santa Claus at the local mall. He initially does well in this job, looking the part, being punctual, amicable, and knowledgeable about reindeer. However, he soon begins to insist on teaching children words like "invertebrate." He also starts giving out stylish feces beads instead of candy canes. Later, he gets in an argument with another Santa Claus in another mall over biologically sound explanations for Rudolph's glowing nose. The "Darwin vs. Santa Claus" fistfight goes viral on YouTube.

WEATHER IS NOT A PEST

With summer past, I remember the flies and other assorted citizens of bugdom at my house. Some were silent like models of mathematical motion, and some buzzed loudly, almost as if you could see their pursed lips – air forced through their invertebrate skeletons. All seemed pervasive, as if to target my children endlessly whilst they play. And I remember my paternal instinct kicking in, deciding that I must do something about these flies. Nasty flies.

So in my efforts to learn more, I came across images of my enemy. Images like this one below:

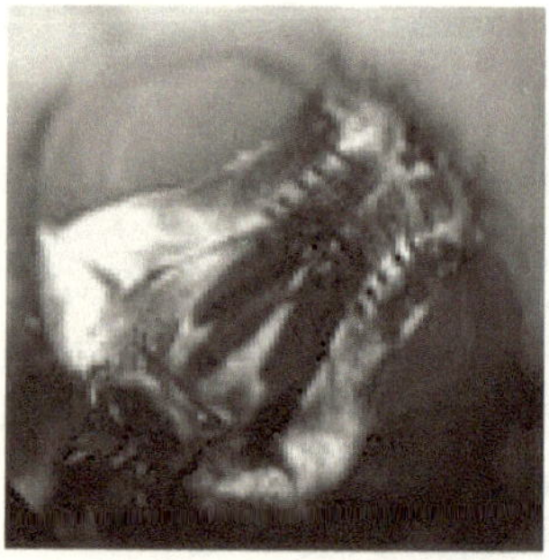

And looking closely, even as a scientist and a follower of the empirical, I could marvel at the inherent beauty of such structures, drawn as if guided by fluid lines, swaths of colours for effect. But it was when I stepped back and glanced distractingly upon the picture, that it hit me – that I've seen this fly before – looking a little cleaner, simplified, but no less the same.

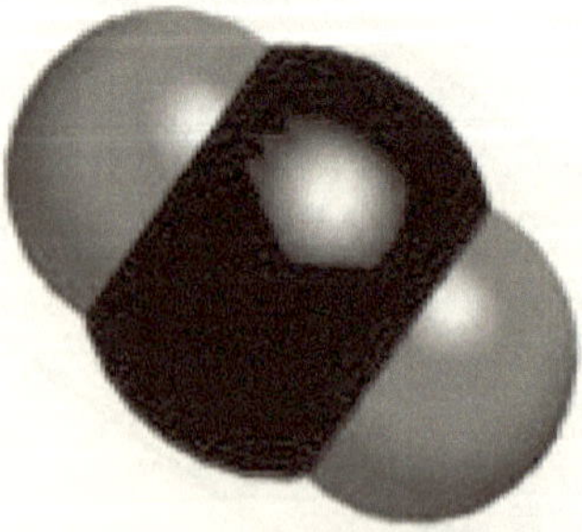

This, by the way, is a picture of carbon dioxide (CO_2 for short) in standard chemical modeling nomenclature (black is carbon, red is oxygen).

Currently, it is estimated that the amount of carbon dioxide in the atmosphere hovers just above 400 parts per million. In comparison, the numbers of flies sharing that same air space is decidedly less so. Both represent scourges of some manner that inhabit the sky. What's interesting to me, apart from the obvious visual similarities, is how the two are perceived in public light.

Of course, people's opinions (my own included) on flies are generally not good. As alluded to earlier, the distain often crosses into the need to remove them altogether. And folks have been ingenuous indeed with the use of traps that use ultraviolet light, pheromones, toxic baits, sprays, surface applications, electric zapping, and (ironically) carbon dioxide. Really, it seems that there is little mercy in this regard.

In contrast, the reaction against carbon dioxide, though apparent, appears otherwise muted and at times in conflict with itself. How else do you explain the wincing over the international climate change agreements, domestic fossil fuel wranglings, or our love affairs with SUVs?

The point, I suppose, is that this is not good. Those 400 parts per million should actually be scaring the shit out of people. And even if you do disagree with things of the global warming nature, there is the reality that one still shouldn't take the chance. Climate can enhance the non-linear effect of weather, and weather, after all, is not a pest.

THE REASON IS MATH

My Fellow Citizens.

I speak to you today on a matter of great importance, a matter that affects the very fabric of our nationhood. You no doubt know that it is the intention of our government to increase our military engagement in any number of hot zones – more bluntly, to send in even more of our fine citizens into the area. Today, I would like to speak to you, our citizens, on why we have chosen such a route. I would like to speak to you on why we feel this is the best route – a hard but worthy choice in a political situation that represents one of our greatest challenges in history.

The reason that this is the best route is simple, my friends. The reason is math.

Now, I know this might trouble the skeptics out there, but the truth is, is that I've done a lot of math in my time. For instance, I know a thing or two about budgets, and budgets are the stuff of numbers. I know that a dime has ten cents, a dollar has one hundred, and that good old Abe Lincoln is worth 500 cents. I also know that when you add those numbers up you get 610. And maybe even more important, is the fact that I know all of these numbers are integers, a word that I am not afraid to use.

Here's another example. Nowadays, I hear the word genetics a lot in all of this stem cell technology stuff. I can tell you that I know that genetics is about the DNA, which is composed of four different nucleotides. Not three, not five, but four. That's also math. Now I might not know what the nucleotides are themselves, but the reality is, my friends, is that that's just not important. That's genetics and stem cell stuff, people – not math. And the reason for sending more troops into Iraq isn't about genetics and stem cell stuff – it's about math.

If I haven't convinced you yet, then you should also know that I've done a lot of math in other aspects of my role as an elected leader. Like the time when my religious friends said that human beings and the Earth were created in seven days, and my scientific friends said, "No sir. This actually took about four and a half billion years."

And you know what I did? I took the mean value between the two – I said, "Let's just say that the length of time is the average between the two, because then you get two and a quarter billion years old, and that's a value I can live with."

I'm also pretty good with graphs. Seriously, I can't tell you how many weather graphs I've seen lately. Graphs about temperature, rainfall, wind speeds, graphs about gases of some sort. But it's all good, and now I even like graphs. Graphs are like math in picture form. That helps because sometimes these graphs have fractions and decimal places, which are much more complicated than integers. Anyway, those graphs are all heading up. Which must be fine, because no-one wants to see a graph heading down, right?

So in the end, my fellow citizens, what I hope I've done is convince you that my credibility in math is sound. And like I said before, math is the best reason to send in more troops. Basically, if you have more good guys than bad guys, then that just works – it's as simple as that.

And you must believe in this logic, because there's even a little special math symbol for things having more than the other – it's called "greater than" by the way, and it looks a little like a sideways "V."

So there you have it: "greater than" has its own little symbol. Do you hear what I'm saying? I'm saying that my reasoning for sending more troops even has its own little math symbol, and really now, you can't argue with that. Besides, isn't that what we aspire to? To be "greater than," integers and all.

ADVICE ON HOW TO BABYPROOF YOUR MOLECULAR GENETICS LABORATORY

One of the first things that a newborn experiences is not necessarily the warmth and scent of the mother's embrace, but rather a series of pokes and pricks to ascertain health and mental alertness. It therefore seems to me that a natural progression of this trend is to incorporate the highest medical predictive technology into an infant's normal surroundings. In other words, it seems obvious to me that sooner or later everyone will have their own molecular genetics lab in their household – most likely adjoining the kitchen.

But, of course, with this new standard of living, steps must be taken to ensure the safety of the child. As a result, I'd like to take a moment and share some of the babyproofing tips that have worked in my household.

- – -

1. Glassware
Thankfully, most of the up to date laboratories rely mostly on sterile plastic ware, so danger due to broken glassware is generally not an issue. As a bonus, your child will likely learn the word "centrifuge" at a remarkably early age.

2. Chemicals
Preferably, all chemicals should be stored in a place that is safely out of reach to prying hands. However, if this is not possible, there should be active steps to label the chemicals according to their hazard level. Color-coding does not work unless the child is at least 2 years of age and capable of identifying colors. In fact, we found that the most effective way of labeling is to adhere Disney characters correlating scariness to relative toxicity. For instance, a picture of Thumper would work well with Sodium Bicarbonate, whereas a picture of scary ass Ursula (from the Little Mermaid) would work well with Arsenic compounds. **WARNING:** do not use pictures of Goofy as infant responses vary greatly.

3. Flammable Reagents

Playing with fire is dangerous at any age, but especially more so in the presence of highly flammable liquids like ethanol and methanol. Although normally kept safely behind the doors of special non-flammable metal cabinets, this is still a problem area since most of your child's fridge magnets will reside here as well. What worked well for us was to take our child's favorite stuffed toy at their earliest impressionable age (about 4 to 6 months), douse it in one of these solutions, and set flame to it. A bit hardcore, but it worked.

4. Radioactive Area

The radioactive area is tricky since it usually incorporates two common pieces of equipment that are extremely attractive to youngsters. These are, of course, the Geiger counter (has a handle, buttons, makes a loud beeping noise, and has a detector that looks remarkably like a microphone), and various sheets of radioactive shielding (great for forts!). My advice is to provide duplicates so that the child can play happily with the non-contaminated versions. **EXTRA TIP:** get a Geiger counter with a mute option – trust me, you will thank me later.

5. Biohazard Area and Disposal

Really now. If you've read the definition of "biohazard" carefully, you'd have already realized that your child's front and back end are part of this category. It's almost as if the whole lab can be your diaper changing area, which in my opinion is wonderfully convenient. Score one for technology.

MY OWN SHORT ILLUSTRIOUS COLLABORATION WITH FRANCIS CRICK

CRICK: Is that your Ford Escort?

ME: Yes it is.

CRICK: It's in my parking spot. Can you move it?

ME: Yes, definitely. Sorry about that.

CRICK: No worries.

- - -

I met Dr. Crick at San Diego's Salk Institute during a summer trip in my graduate student days – although "met" is perhaps a verb with too much significance in this case. I was actually there to touch base with some old friends of mine and was told to park in his spot since we would only be 15 minutes or so. In truth, we were en route to Anaheim, Disneyland specifically, and bumping into scientific legends was the last thing on our minds.

Dr Crick, of course, is well known for his discoveries in the world of DNA, being one of the individuals responsible for figuring out how the A, T, C and G's of genetic code stacked up. But later in life, he took an interest into the mysteries of consciousness. In particular, he was intrigued at how the brain so quickly generates visual awareness upon viewing a scene (or something like that). It's an interesting biological question, in that I know I'm curious to understand what goes on when you look upon the world – or perhaps in more profound instances, what happens when a child first sees the Magic Kingdom, when a soldier stares down the barrel of a gun, or when you first meet the person with whim you will, unbeknownst to you, fall in love with.

Almost the minute we parked our Ford Escort, Dr. Crick pulled up in a large stately white car, a Mercedes or a Cadillac I think. He got out, dressed I can only describe in a manner that approximated most perfectly his vehicle, and politely asked that I move. I obliged immediately. Looking back, I often wondered what his consciousness was telling him when he saw me that day. It's probably quite different from what my own brain was experiencing: I just thought it was cool that his license plate read "ATCG."

IT'S A LUCKY THING FOR EVOLUTIONARY BIOLOGY THAT THE FOLLOWING PASSAGES AREN'T IN THE BIBLE

Jesus then entered the farm, and saw creatures of every shape and size, and so said to his followers, "Hey, my Dad made that creature, and that creature, and also that creature… Actually, now that I think about it, he made them all."

And at the early dawn of the seventh day, just before He rested, God did a lot of pretty complicated things at super duper God speed. This was so that people would think the whole Creation thing probably took a lot longer than seven days.

The heavens opened and the angels proclaimed, "Fear any literate man, capable of impressive facial hair, who is comfortable on boats, has a thing for finches, and is named Darwin, for he is basically an unrighteous phony. So it is said in the very literal Kingdom of God."

And the Lord said, "Yes, my child, the unicorn was a first edit. They were poorly designed so I had to do away with them – kept goring themselves when nuzzling and stuff. Indeed, not my best work."

With Cain facing Abel, God then commanded, "Look deeply into his eye, and marvel at my handiwork, because my child, making that eye work properly, took, like, for freakin' ever!"

And God appeared to Moses as a Burning Bush – not a monkey, but a bush. Because clearly, God is no monkey.

NONE OF MY SCIENCE PIÑATAS ARE APPROPRIATE FOR CHILDREN

1.
Hydrochloric-Acid-Filled Piñatas

Good: Have the sturdy construction required to ensure no unintended leakage of contents.

Bad: Possible severe burning. Brings the party down.

2.
Endangered-Animal Piñatas

Good: Kids love animals. High potential for very cute-looking piñatas, like baby seals, for instance.

Bad: Beating with a stick sort of sends the wrong message.

3.
Particle-Accelerator Piñatas

Good: Built full-scale and often several miles in dimension. Therefore, young children find them easy to hit.

Bad: Each one worth several billion dollars. Parents generally not keen on damaging them.

4.
Smallpox (Variola major) Piñatas

Good: Cool virus shape.

Bad: Highly contagious and high mortality rate. Would also bring party down—as well as everyone else within a 100-mile radius.

5.
Infinity-Symbol Piñatas

Good: Possibly a way to address the often reported decline of mathematics education.

Bad: Thinking about infinity makes my head hurt. Now imagine having to explain it to a child over and over again.

6.
Piñatas in the Shape of the USA and Filled With the Greenhouse Gas Carbon Dioxide

Good: Sort of work as a metaphor for the United States' role in the global-warming crisis.

Bad: Unfortunately, the irony would be totally wasted on a 5-year-old.

LIKELY AND UNLIKELY THINGS THAT SIR ISAAC NEWTON STOOD ON DURING HIS LIFETIME

Likely:

Grass.
A stage of some sort.
Guard.
Tippy toes.

Unlikely

Astroturf.
Olympic podium.
Someone's throat.
Shoulders of (Real) Giants.

THE BESTEST, MOST KICK ASS, HUMAN GENOME PROJECT

Mondo-Genetic-Services is proud to announce its latest venture, "The Bestest, Most Kick Ass, Human Genome Project." Hot on the tails of the International Human Genome Sequencing Consortium and Celera Genomics, we present to you a novel approach in the elucidation of mankind's blueprint of life. Rather than using the frequently studied yet boring human cell lines, or samples from a small group of ethnically diverse, anonymous, and likely dull individuals, we propose a completely different strategy – that is, we plan to use the genomes of individuals handpicked by the editorial staff of People magazine, a move we feel will cater to the desires of you and your friends. Currently our impressive roster of prospective subjects include the following:

People's Choice Favourite Motion Picture Actor – Harrison Ford
How can any human genome project not have samples from the man revered as Han Solo and Indiana Jones? The man who has uttered such immortal words as "Punch it Chewie," and "Nazi's – I hate these guys." In related news, Mondo-Genetic-Services has also tried to recruit his girlfriend Calista Flockhart into the project, but has recently learnt that she simply did not have enough tissue.

People's Choice Favourite Motion Picture Actress – Sandra Bullock
Mondo-Genetic-Services feels that the inclusion of Ms. Bullock, the purveyor of such classics as Speed 2 and Miss Congeniality, into the Bestest, Most Kick Ass, Human Genome Project is practically self explanatory. Because really? Who doesn't like Sandra Bullock?

People's Choice's Favourite Performer in a Children's Television Program – Goofy
Is he a man? Is he a dog? Is he a man-dog? Be one of the first to find out, here at the Bestest, Most Kick Ass, Human Genome Project.

People's Choice Most Interesting Person in Politics – Olusegun Obasanjo
Through email correspondence, the editorial staff of People Magazine have finalized an agreement to sequence the DNA of past President Obasanjo, of

Nigeria. In return and given their capacity to act as an overseas partner in a balance account transfer from the Central Bank of Nigeria, he will place 20% of US$21,320,000.00 (TWENTY ONE MILLION, THREE HUNDRED AND TWENTY THOUSAND U.S DOLLARS) into their corporate accounts.

People's Choice Most Interesting Person in Sports – Michelle Kwan
Yes, the folks at People magazine are certified KWAN FANS. Michelle has agreed to participate in this project and in return, we will help start up an official Michelle Kwan fan club. More to the point, inclusion of DNA from this outstanding athlete will allow us to finally answer one of life's most troubling questions – that is, how exactly does figure skating get judged?

People's Choice Favourite Television Icon – Arthur Fonzarelli
"The Fonz" was a cultural icon of the 1950's and is certainly deserving of a place in the Bestest, Most Kick Ass, Human Genome Project. Not only did he seem to have telekinetic powers, but this is one guy who must have seen a lot of sex! Since the lubricated condom wasn't introduced until 1957, and the oral contraceptive wasn't even invented until the 60s, Mondo-Genetic-Services wouldn't be surprised if Mr. Fonzarelli himself sired half of Middle America.

People's Choice Favourite 80's Television Comedy Series – Cast of "Who's the Boss"
In an attempt to secure DNA sequences that espouse the best of American family virtues, the Bestest, Most Kick Ass, Human Genome Project will obtain tissue samples from the entire cast of "Who's the Boss." This will include cells taken from Tony Danza, Judith Light, Katherine Helmond, Alyssa Milano, and even the little boy whose name no one can remember.

People's Choice Favourite Diety – Jesus:
In a coup d'etat for this project, Mondo-Genetic-Services has secured the sole rights to sequence and publish the Prince of Peace's very own DNA. Furthermore, our scientists have also discovered that due to the principle of the Holy Trinity, this agreement also effectively grants us sole rights to the genetic code of the Holy Spirit and of God himself

People's Choice Reader's Pick – George W. Bush
Because apparently America, like the rest of the world, is wondering "what the hell is up with that?"

I SUSPECT THE I.P.C.C. REPORT MIGHT BE MORE EFFECTIVE IF IT WENT WITH ACRONYMS THAT WERE MORE NARRATIVE IN NATURE

The IPCC report[1]

The STWBTIPCCFARCC2013TPSB report[2]

The OWCGWIPCCFR report[3]

The YIACCAYII report[4]

The OKSILTRIWSHAGTATTSAAEOTSOCC report[5]

The BTESCCIRINAGTAAIPOF report[6]

The OKTIA95LTIOFBTB100ITKOATIPITSR report[7]

The IOWSAACASCBAT report[8]

The MYCSTW100CITPRBWAKHTKOSTTTO report[9]

The SIATITRBYKSIW report[10]

The FFSJRTGROALTTRACNPOI report[11]

The ABCIDMOLPCOAGWFDOEICFFIIFFATL report[12]

The SCTINFSC report[13]

The AAYWWMLBGOTKOSMTIIBOLWLOMMOTOACISFMAOGEOCI report[14]

The ASTTQIAYOBWWTIPCCIS report[15]

The BIYTWTFJGALBB report[16]

The INWTTARSIIFHSLCHUMCPBTLUIPHBSWGTGWA95CTYCAYCCAGTBMTALDWY report[17]

- - -

1. Intergovernmental Panel on Climate Change

2. Specifically, this would be the Intergovernmental Panel on Climate Change, Fifth Assessment Report Climate Change 2013: The Physical Science Basis

3. Or we could go with Intergovernmental Panel on Climate Change For Realz

4. Yes, it's about climate change, and yes, it's important.

5. O.K. So it's like this report is where several hundred academics get together and try to summarize all available evidence on the science of climate change.

6. Basically the evidence says: climate change is real, it's not a good thing; and also, it's partly our fault.

7. O.K. Technically, it's a 95% likelihood that it's our fault, but that's because 100% is the kind of assessment that isn't possible in the scientific realm.

8. In other words: scientists are as certain as scientists can be about this.

9. Maybe you can say things with 100% certainty in the political realm, but we all know how those kinds of statements tend to turn out.

10. Seriously, it's all there in the report. Because, you know… Science, it WORKS.

11. For fuck's sake, just read the goddam report! Or at least try to read a credible news piece on it.

12. And by credible, I don't mean outlets, lobbyists, political commentary or advocacy groups where funding directly or even indirectly comes from folks invested in fossil fuels and the like.

13. Scientific conspiracy? There is no fucking scientific conspiracy.

14. Also, ask yourself: who would most likely be guilty of that kind of spin? Messaging that is influenced by oil lobbyists with lots of marketing money? Or thousands of academics conspiring in secret faculty meetings and organizing grand exchanges of covert information?

15. Anyway, screw this. The question is, are you on board with what the IPCC is saying?

16. Because if yes, then wonderful! The future just got a little bit better...

17. If no? Well then, that's a real shame. Isn't it funny how scientific laws can help us make climatology predictions, but they're less useful in predicting human behavior? Still, we're going to go with a 95% certainty that your children and your children's children are going to be more than a little disappointed with you.

ANGRY WORDS FROM A GNOME WHO TO THIS DAY CONTINUES TO THINK THE HUMAN *GENOME* PROJECT WAS ACTUALLY THE HUMAN *GNOME* PROJECT

It's hard to believe that 17 years ago the Human Gnome Project formally began. It was quite frankly a great day for all of us gnomes as we thought we had finally gained the attention and respect we deserved as a community. But 17 years later, we as a community are disappointed, angry, full of resentment, and still addicted to nicotine.

To our knowledge, of the roughly $3 billion worth of research funds given to the human gnome initiative, none of it ever actually went to fund "gnome" research. Instead, a sizable portion went to human research, and in an apparent slap in the face to my kindred, significant amounts also went towards research looking at bacterial, yeast, worm, fly, and mouse genetics. Suffice to say, that with the exception of humans, these are all organisms that do not smoke. To say that this has been hard on my community is an understatement of vast proportions. Apart from the soaring lung-cancer rates, I find I am continually aware of other lost opportunities the money could have been used towards.

For instance, for whatever reason, we as a race are forever doomed by our incessant need to wear pointy hats. I hate my stupid hat—loath it with a passion. And yet I have to wear it. We all do. Why this is so has been mystery for many an age. Maybe that's why I go through 70 grams of tobacco each day. And whilst pointy hats are fine for garden work (one of our main sources of economic recovery), they are hardly advantageous in the current global market—especially when first impressions play a key role. Surely, there is an underlying neurological basis for this behavior—a basis that science could have elucidated.

And what about our facial hair? Believe me, it is not because we are particularly fond of our beards. It's not even because tobacco pipes look cooler in this context. Our beards just happen to grow at amazingly fast rates! This is not such a huge issue with me and the other male gnomes, but my poor wife actually has to shave every 45 minutes or else deal with social harassment (although if you ask me, I enjoy her hairy legs or armpits). This is also compounded by the fact that services, like laser hair removal or electrolysis, are just too expensive, especially on a gardener's income.

Ironically, the only gnomes who could possibly afford these high tech solutions are the few who have made it into Hollywood where maintaining the typecast "bearded" look is required anyway. Furthermore, even when a hairless gnome is needed on a movie set we still get passed over because of our goddamn pointy hats! I bet $3 billion could have sorted this problem out a long time ago.

But if there was ever a strong case for gnome research, you only need to look at my poor Uncle Bill. This unlucky bastard of a gnome must have some bladder problem or something, since he is (no exaggeration) urinating *constantly*. Seriously, I don't think he's even had a chance to put his penis away since he started 14 years ago! And the truth of the matter is that this particular problem is relatively rampant in my circles. Most start off fishing, or maybe pushing an empty wheelbarrow, and then they feel the urge and then whammo! It's like a disease. I don't think it's too difficult to appreciate the magnitude of this medical condition. Aside from the psychological pain endured, imagine how uncomfortable it must be to leave it "out" constantly in all manner of weather conditions. I don't care if you are the gardener type— when it's cold, it's cold! Plus, it makes smoking a pipe tricky.

Anyway, I'm not here to preach endlessly about our problems. I just here to say I want a fair piece of the action. If the project is called the Human Gnome Project, then it only makes sense that at least *some* of the money should go towards gnome research—right?

O.K., I've said my piece. I really have to go outside now to smoke my pipe—stupid human nicotine patch, piece-of-crap waste of money

COMMON SAYINGS TRANSLATED INTO SCIENTIFICALLY CORRECT STATEMENTS

COMMON: Hit the ground running.
SCIENTIFIC: Conserve momentum.

COMMON: My gut was telling me.
SCIENTIFIC: My colon speaks.

COMMON: If I had a nickel for every…
SCIENTIFIC: x times 5cents, where x equals…

COMMON: Don't throw the baby out with the bathwater
SCIENTIFIC: Avoid wet baby head trauma.

COMMON: Go the whole 9 yards.
SCIENTIFIC: Displace by 8.2296 meters.

COMMON: You are in way over your head.
SCIENTIFIC: Anatomically speaking, you are likely upside down.

COMMON: You spilled the beans.
SCIENTIFIC: Entropy went up.

COMMON: A little bird told me…
SCIENTIFIC: Whilst under the influence of psychedelic hallucinogens…

COMMON: Everything but the kitchen sink.
SCIENTIFIC: Almost, but not quite, the entire universe.

COMMON: Beating a dead horse.
SCIENTIFIC: Technically still dead.

COMMON: Break the ice.
SCIENTIFIC: Break the ice.

THE ABRAMS' STORMTROOPER AXIOM

It (hypothetically*) goes like this:

$$(1) \quad \frac{W_k S_{blaster} \cdot bmi_{opt} \cdot h_{opt}}{5.4^{(1 + b_p^o + b_w^o)}}$$

$$(2) \quad bmi_{opt} = \left[1 - \left(|(25 - mh^{-2})| \cdot 0.04\right)\right]$$

$$(3) \quad h_{opt} = \left[1 - \left(|(h - 1.8)| \cdot 1.112\right)\right]$$

$$(4) \quad \frac{W_k S_{blaster} \cdot \left[1 - \left(|(25 - mh^{-2})| \cdot 0.04\right)\right] \cdot \left[1 - \left(|(h - 1.8)| \cdot 1.112\right)\right]}{5.4^{(1 + b_p^o + b_w^o)}}$$

* Like all good science, this needs some testing…

When news hit that Disney bought the rights to Star Wars, and that J.J. Abrams would be manning the first movie of a new trilogy, my inner geek went into giddy overdrive. This was because it gave me a chance to revisited my bucket list, which had previously scratched off "be an extra in a Star Wars movie" as something that was unattainable having presumed the prequels were my last chance. But now, there is (literally), A NEW HOPE. Even better, is the fact that my kids are old enough to also want this.

And so, being a science-y sort and all, I figured the first step would be to actually try and come up with a way to calculate the odds of such a thing happening, and hence you see the above – or what I have termed the **Abrams' Stormtrooper Axiom**. In effect, this is an equation that aims to calculates the odds of you (or anyone) being cast as a stormtrooper in one of these new movies[23].

---

[23] In general, I've used information from the original trilogy for points of reference.

Here's how it works. We'll first look at (1) which expresses the equation in its most obvious form.

$$\text{(1)} \quad \frac{W_k S_{blaster} \cdot bmi_{opt} \cdot h_{opt}}{5.4^{(1+b^o_p+b^o_w)}}$$

When you look at this equation, there are three main components: two in the numerator: $\boldsymbol{W_k S_{blaster}}$ and $\boldsymbol{bmi_{opt} h_{opt}}$

And one in the denominator: $\boldsymbol{5.4^\wedge(1+b^o_p+b^o_w)}$.

The denominator is an expression designed to address the likelihood of being cast, as having a dependence on the individual's chance of contact with J.J. Abrams. Specifically, $\mathbf{b^o_p}$ refers to the degrees of personal separation the individual is from the Director, whereas $\mathbf{b^o_w}$ refers to the degrees of internet separation the individual is from the Director. The base of the exponential relationship is, of course, the standard ***May The Force Be With You Constant*** (or **5.4**).

All told, if you have very little connection to the director, your odds can dwindle significantly, about $\mathbf{5.4^{(1+6+6)}}$ times, or roughly one in **3.3 billion**! It also infers that even if you know JJ Abrams personally, it does not guarantee being cast – mathematically, the closest association would still work out to $\mathbf{5.4^{(1+1+1)}}$, or roughly a chance of **one in 158**. This is because there are other factors that need to come into play when determining whether an individual is right for a stormtrooper part.

Which is where the numerator expressions exert their influence. We can first begin with the $\boldsymbol{bmi_{opt} h_{opt}}$ element, which essentially considers the physicality of the individual vying for a stormtrooper part. The ***bmi*** portion considers body shape, whereas the ***h*** portion considers height.

Each element can be further derived as:

$$\text{(2)} \quad bmi_{opt} = \left[1 - \left(|(25 - mh^{-2})| \cdot 0.04\right)\right]$$

$$\text{(3)} \quad h_{opt} = \left[1 - \left(|(h - 1.8)| \cdot 1.112\right)\right]$$

Where **(2)** calculates divergence from an average body type (as expressed by an individual's body mass index with $\boldsymbol{m}$ equals to the individual's weight in **kilograms**, and $\boldsymbol{h}$ is equal to the individual's height in **metres**). You'll note that the more you veer away from an "average" body type, the greater the modification of the $\boldsymbol{bmi_{opt}}$ number to a number less than one (and therefore further lowering your odds).

In the same manner, **(3)** calculates divergence from an optimal height (deemed **1.8 metres** as determined from casual examination of Star Wars' trivia – i.e. calculating Mark Hamill's height and noting the "Aren't you a little short to be a Stormtrooper?" comment). Like the BMI calculation, the more you deviate from the optimal height, the greater the modification of the $\boldsymbol{h_{opt}}$ number to a number less than one (and therefore further lowering your odds).

Note that both **(2)** and **(3)** are included in the overall equation for pragmatic prop design reasons (not every extra can have a custom made set of armour, so it makes sense if casting aimed for similar body types). Then, of course, there is the whole clone army narrative which might also presume the troops having similar physical features. (Also note that in case you weren't familiar with the symbol, the straight up and down lines enclose a value where you only use the absolute number – i.e. remove the plus or minus sign).

Anyway, when you put it all together you get the expression (4).

$$\frac{W_k S_{blaster} \cdot \left[1 - \left(|(25 - mh^{-2})| \cdot 0.04\right)\right] \cdot \left[1 - \left(|(h - 1.8)| \cdot 1.112\right)\right]}{5.4^{(1 + b_p^o + b_w^o)}} \tag{4}$$

Which only leaves $\boldsymbol{W_k}$ and $\boldsymbol{S_{blaster}}$ to be defined. Here, these two variables relate to two specific personality traits that are deemed important for the stormtrooper casting decision.

For instance, I don't think I'm the only Star Wars fan who notices the incredibly poor marksmanship exhibited by the stormtroopers. There are many instances in the movies where there are many of them (with their weapons – presumably high tech in nature), in close proximity to the target, and yet, they still always fail to hit their target.

Given this observation, I'm left to assume that Stormtroopers, as a whole, have a deep distrust of guns, and with that discomfort tend to misfire (perhaps subconsciously). This also leads me to hypothesize that not only are they not very skilled, but that they are probably the sort that are not at all familiar with gun culture in their private lives.

Consequently, $\boldsymbol{S_{blaster}}$ is a number assigned to measure the individual's relative experience with guns, whereby a value of **1.0** represents full disconnect from the use of guns in their personal lives, and a number closer to **zero** represents an individual who is very familiar with gun culture.

Of course, perhaps the most important tangible characteristic (that could translate to a positive casting decision) is relative fandom itself. In other words, casting may be partly governed by how "*into* Star Wars" an individual is. Here, and in honor of Chewbacca's reference of "pulling arms out of their sockets when they lose," I've decided to use **Wookie knowledge**, or $\boldsymbol{W_k}$ as an indicator that can further increase casting chances. Essentially, this is a scale that ranges from **1** to **10**, whereby **10** represents fanatical knowledge on all things Wookie, and **1** represents no knowledge at all. In effect, if you're nuts about Star Wars (and wookies specifically), you can increase your chances of being cast by 10 fold.

In conclusion, I want to stress that this is the **Abrams' Stormtrooper *Axiom***, and by its very definition, an *axiom* is just a starting point. This means the equation will need more work, and it would be great suggestions to make it better. As it stands, it works as a general guideline using a number of test values[24]. As well, there is also the very real caveat of whether J.J. Abrams will even have stormtroopers in the new movies – never mind the fact that if he does, they may come in a different size, or be better at shooting, etc. In some respects, this reminds me a little of Schrödinger's cat (we can call our version Abrams' Stormtrooper): we won't really know what he has in mind until he lets us open the box.

---

[24] For instance, an individual with no connection at all will result in a number that works against the backdrop of the total human population numbers. For J.J. Abrams, himself, where $\mathbf{b^0_p}$ and $\mathbf{b^0_w}$ are equal to zero, and his W_k is likely quite high, the equation would further calculate that he has practically perfect odds of being cast as a stormtrooper (which makes sense given his role in the movie). For the sake of comparison, I've calculated my own odds to be approximately:**0.00000519** or about **one in 19,000**.

A FEW WORDS FROM THE SCIENTIST WHO INVENTED WONDER WOMAN'S INVISIBLE JET

First of all, I totally get it. You're watching *Super Friends* or reading some *Justice League of America* comic book, and what do you see? Wonder Woman floating in mid air, in what is apparently an "invisible jet." And for some reason, the fact that *you* know that it's an invisible jet (because someone has gone to the trouble of outlining the jet with white phantom lines), is supposedly meant to make this OK.

Except that it doesn't. You know this and I know this, and well, everybody knows this. Because the truth is: Wonder Woman looks kind of stupid floating in mid air. I mean, seriously, what is the point of having an invisible jet if the pilot—and a pilot wearing bright sparkly superhero colors—is so… well… visible?

So I get all the hate, I totally do.

But listen: I *invented* that invisible jet! That invisible jet is *my* research. It's *my* baby.

And the reality is that the invisible jet was never meant to *hide* the pilot. It was just meant to be invisible on its own. If you don't believe me, then by all means, look it up—my doctoral dissertation and my research publications are quite clear about this.

Wonder Woman totally got this. Actually, she was pretty amazing about it. She was like, "whoa… invisible jet… that's pretty cool." And then I was all like, "Yeah, but you know that it's *only* the jet that goes invisible right? You know that *you* won't be invisible when you sit inside it, right?" And she was like, "Yeah, I get it, but that's OK, because well… *invisible-freaking-jet*!"

She basically said that it was totally fine to spend a whole bunch of her money on it because (a), it was a good way to support interesting science, and (b), well… it was just awesome.

Unfortunately, her Super Friends buddies were all idiots about it. They were all like, "Diana… you know we can still see you?" And Superman was prone to flying in a seated position and making steering motions and going, "look at me,.. who am I?" Meanwhile, Batman was all like "my utility belt is way cooler."

But she knew what was going on. She knew that the reality was this: that the jet is a goddamn SCIENTIFIC MARVEL! Something that should be admired—because ingenuity, years of hard work, and significant research funds was all involved to develop that piece of technology.

Now, do I want to continue my research so that things entering the jet also become invisible? Sure—because that would also be pretty awesome. And maybe, one day, I *will* work on that research goal, and if I do, Wonder Woman is totally going to get first dibs on that piece of hardware.

Because listen, folks: this is how science works, one small step at a time. It's not fast (like a speeding bullet), or able to make progress (in single bounds). It's not about gimmicky quick fixes (like the kind you pack in a utility belt), and it takes serious investment (we're not all caped billionaires with money for caves and batmobiles). This discovery stuff is slow and incremental, and that my friends, is the honest to goodness Lasso of Truth.

WAYS POLITICIANS AND ROBOTS ARE ALIKE

Their eyes.

A politician's eyes appear to be fully capable of eliciting an empty yet intense robot gaze. Intriguing because this look is reminiscent of the kind that any human might make when solving difficult math problems in their heads, and which, coincidentally, is also an activity that robots are notorious for doing all the time. On the other hand, maybe they look that way (the politicians) because their eyeballs are also recharging and getting ready to shoot out laser beams.

Their focus.

Politicians are always super focused and always "on message," sometimes to the point of "muzzling" individuals who might veer away from their specific agenda. This, of course, is indicative of algorithmic behavior and also of spyware filtering, which when taken to certain extremes is closely associated with "programming" for robots of the evil genius ilk. Indeed, this observation is quite striking: there is an eerie similarity between most Democrats and Dave from *2001*, as well as most Republicans and Megatron from *The Transformers*.

Their message.

Robots, like computers, are often relentless in producing endless streams of spam. As well, this spam almost always fits in one of two categories: (1) either promises for financial wealth and economic prosperity; or (2) pornographic photos of genitalia. Sound familiar?

Their apathy towards unusual climate patterns.

Typically, politicians have a poor record on climate change policy. It is almost as if they don't care that it's happening. Which begs the question:

why the nonchalance attitude? Wouldn't most leaders in our society be wary of what is arguably the single greatest challenge facing humanity today? Is it because they know that they as robots are generally impervious to temperature and weather extremes?

Their wariness of appearing too robot-like.

This particular attribute is most likely to manifest itself as a collection of exaggerated attempts to draw attention away from their robot ways. For example – Possible exaggerated attempt #1: kissing babies. Possible exaggerated attempt #2: serving customers at a small local business. Possible exaggerated attempt #3: saying something very very stupid. Interesting to note that the combined symbolic aesthetic of baby kissing plus humbly serving customers plus saying very very stupid things is widely recognized as the perfect antonym to the "essence of robotness."

Their hearts.

There are some politicians who seem programmed to care little for social programs and/or initiatives that aim to help the less fortunate. This suggests that the concept of inequity is perhaps too difficult to compute. If so, that's a little cold… maybe even *robot* cold.

SOMEWHAT RANDOM ASSORTMENT OF ORGANISMS OR SOMEWHAT INAPPROPRIATE ANIMAL NAMES FOR BROWNIE GIRL GUIDE LEADERS

Barracuda

Fox

Tit

Minx

Cougar

DEAR OPRAH: SOME THOUGHTS ON YOUR CREDIBILITY. AN OPEN LETTER

Now that we're half way through the university semester, I'm finding myself inundated with a lot of marking. Sometimes, I try to tackle this work at home, but being the skilled procrastinator that I am, this will inadvertently lead me into the land of daytime television. It was here the other day that I caught a few minutes of Oprah, and noted that in that short timeframe, I found my reaction changing from a sort of admiration to a feeling best described as a prolonged wince. The reason for this abrupt change of heart was essentially the appearance of Jenny McCarthy in what looked like a correspondence role – she of the celebrity ilk, noteworthy for being a very powerful advocate of some very shaky medical advice.

I won't go into too much detail here about her travails, since they've been covered extensively here at Boingboing and elsewhere in the media, but suffice to say, both the medical and scientific communities overwhelmingly take issue with her claims regarding linkage between the MMR vaccine and Autism. Indeed, her opinion has not changed, despite recent studies that showed that much of the data in the Wakefield paper (the scientific article that laid the media groundwork for this linkage) was actually fraudulent in nature[25].

Anyway, this is interesting to me. Ms. Winfrey by all accounts seems to have her heart in the right place, and as a person of considerable media clout, you would think that she (or at least her team) would have carefully thought through the ramifications of associating with such a notorious individual. Except that when you look a bit deeper, you find other instances where her brand chooses to ignore a very simple and sensible idea: that "claims," especially claims that operate best under scientific ways of knowing, should only be supported when there is robust evidence to back them up.

---

[25] How the Case Against the MMR Vaccine was Fixed. *BMJ* 2011; 342 doi: http://dx.doi.org/10.1136/bmj.c5347

An obvious example of this is her recommendation of The Secret. This is a book written by Rhonda Byrnes and which appears to be a very elaborate and (if I can be cynical here) lucrative interpretation of the placebo effect. Specifically, the author claims that an individual can "change their electromagnetic frequency," so as to change outcomes in their life. Such language is striking because if you were to ask an expert who knows a thing or two about electromagnetic radiation – say a physicist – you would learn that this phrase is entirely nonsensical. More importantly, you could even ask physicists of different moral leanings, political sensibilities, and/or cultural backgrounds, and you would still get the same answer – because the evidence that refutes her claims stands on its own objective merits.

We could go on with other examples of Ms. Winfrey's fondness of pseudoscientific trends – from the establishment of Dr. Oz, to providing the center stage to individuals like Susanne Sommers and Christiane Northrup – but I think you get the point. Let me also be clear: I do think there is some value to these things if individuals truly feel that they are benefiting from them. However, what's worrying to me is when lines regarding safety are being crossed. All to say that for me, there's a bit of irony here, because before seeing Ms. McCarthy on her show, one of things I applauded Oprah Winfrey for was her work in South Africa, particularly her involvement on the HIV/AIDS front.

As many already know, this is a country that continues to be devastated by the effects of this disease. According to the latest UNAIDS statistics (based on 2009 data[26]), South Africa currently has the highest infection numbers, estimated at 5.6 million of its population. This includes a startling 17.8% prevalence in individuals aged between 15 and 49. It's also no secret that a significant part of this deadly reality is due to poor government policy, whereby from 1999 to 2008, the former President Thabo Mebki and his health minister, Manto Tshabalala-Msimang were willing advocates of a variety of pseudoscientific claims made by AIDS denialists. Many of these deterred the provision of life-saving antiretroviral medicines: most infamously, Manto herself promoted the use of "beetroot and garlic consumption" as an effective treatment regime.

This narrative is strikingly similar to those that allude to Ms. McCarthy or The Secret. The difference, of course, is that with HIV/AIDS in South

---

[26] You can check out the 2010 figures at http://www.unaids.org/globalreport/Global_report.htm

Africa, Ms. Winfrey chose to side with reason, data, and good evidence. More to the point: having both Ms. McCarthy and the South African HIV/AIDS issue being so prominent under a single brand is an odd dichotomy that begs us to wonder what to make of it. It is, quite simply, a mixed message. At best, it is confusing in a world where the glut of information is already a burden. And more seriously, it is an insult to the good people who have worked so hard on HIV/AIDS education, treatment, and research. But at its worst, it is an affront to all those who have been victims of the propagation of such dangerous claims, whether it is the people of South Africa or the millions of viewers that follow Ms. Winfrey's every suggestion, every recommendation, and every action.

TYPES OF SHARKS THAT ALSO SOUND LIKE HEAVY METAL BANDS

Great White

Clouded Angel

Hammerhead

Thresher

Leopard

Megamouth

Requiem

THE CROCODILE HUNTER BECOMES THE PLANET HUNTER

Dedicated to the man who was one of the funniest things, ever, on television, and we mean that in only the best of ways

Cor Crikey! And g'day mate! Right now we're walking up to Hawaii's Gemini Observatory on the summit of Mauna Kea. It's got a beaut of a telescope inside, and we're hoping to find a new planet today.

(Whispering) Here we are at the front door. But we should first give it a bit of space. Patience is important when dealing with telescopes. And we've got to be careful with that door. It's locked! Looks like the observatory doesn't open for another 20 minutes.

(20 minutes later) Alright mate! Let's go!*(running)* Quickly mate! We're already inside, but we've got to move fast! If you look around, you might see that there are other humans around here that will also want to use the telescope, but if you get there first, you're in there mate. You can use one hand for the controls, and the other to fend the others off.

(Reaching the console) We're the first here! And it looks like we'll get to have it to ourselves too. Ripper! Looks pretty complicated, but I've been around telescopes all my life and this is definitely an "on" button. But before I press it, let's first camouflage ourselves behind this adjustable office chair, just in case! I'm going to turn it on now.

(Apparatus makes a noise). Watch out mate! We've got to stay extra alert now. Remember – never do this without the supervision of an expert like myself around.

It's on. And don't forget to be on the look-out for other humans. We can scare them off by making ourselves look as big as possible – spread your arms wide and look like you're real pissed. That's right, like that. Beauty mate! Alright, now let's go find us some planets…

(7 hours) Did you see that?

(12 days) Did you see that?

(4 weeks) Did you see that?

(6 weeks) Did you see that?

(7 weeks) Crikey! Did you see that?

(3 months and 1 week) Did you see that?

(4 months) Did you see that?

(5 months and 3 weeks) Did you see that?

(6 months later and looking weary) Well mates, that's all we have time for in this show. It's a shame we didn't find a new planet but that's sometime how it is in these observatories. See you next time!

IT'S A LUCKY THING FOR STEM CELL RESEARCH THAT THE FOLLOWING PASSAGES AREN'T IN THE BIBLE

The petri plate is the work of Satan. How does God know what a petri plate is in this ancient time before the advent of scientific achievement? It is because he's God, which is really handy for that sort of thing.

Go forth my children and use the word "embryo" whenever you can. It is a very pleasant-sounding word—say it as often as possible. In fact, my children, try this: point to anything and everything and say, "That's an embryo."

Mary tells us, "When a sperm and an egg come together, it represents the ultimate act of compassion and love. Therefore, it is a grievous sin to do studies on this type of thing. Plus, it's also kind of private."

The Lord says that our precious hearts and minds represent flesh of enormous piety. They should never be regenerated, regardless of the circumstances. While we're at it, we should also never regenerate eyebrows, nosehairs, or nipples—although the Lord figures that that is a given anyway.

For people who have had an accident and have lost the use of their legs, it is not the way of the Lord to try to fix this pain. Instead, God will tell them, "That's too bad." Then he will likely tell them a good joke to make them feel better.

And Jesus said, "Liquid nitrogen is evil. Once, while playing with it, I froze my finger solid and it actually broke off. Lucky for me I'm the Son of God, and I can just grow another one."

BREAKFAST OF CHAMPIONS DOES REPLICATION

To begin with, we'll start with a chicken scratch drawing of a DNA molecule, which you know is double stranded. My poor pathetic attempt at illustration is therefore going to look like this:

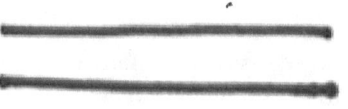

You also know that each strand of DNA is composed of building blocks called nucleotides, and that these nucleotides are always interacting in a complementary manner. For example, A's are always with T's, C's are always with G's, Beavis is always with Butthead, etc etc etc. Let's draw them in like so:

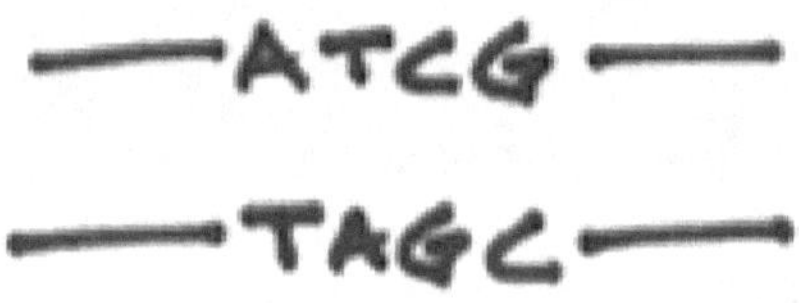

What you haven't been told at this point is that chemically speaking, the two strands are going in opposite directions. The correct term for this is actually known as *anti-parallelism*. To denote this, I'll draw some arrowheads on the DNA strands:

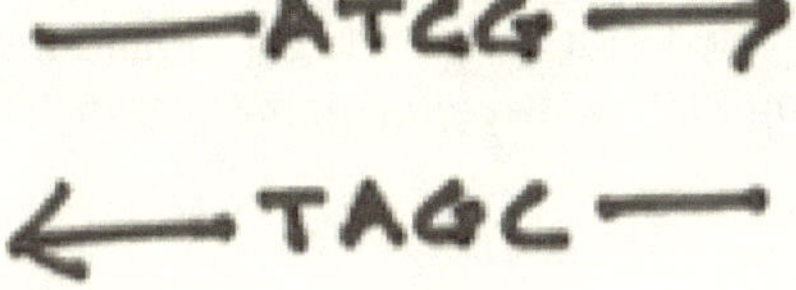

Although, this may seem a little confusing at first, try to picture two lines of square dancers facing each other. In this circumstance, you notice that when focusing on the left or right hands of the row of dancers, the two lines are going in opposite directions. This picture should help:

Your DNA strands are doing something very similar in a chemical sense. The difference, of course, is that instead of dancers, you have your choice of four nucleotides. Furthermore, like the situation of left hands versus right hands, the ends of the DNA strands are also different. One end is known as the 3' (pronounced 3 prime) end and the other is known as the 5' end. To the layman, these rather stoic terms are an unfortunate consequence of chemical labeling. So now, our picture should look like this:

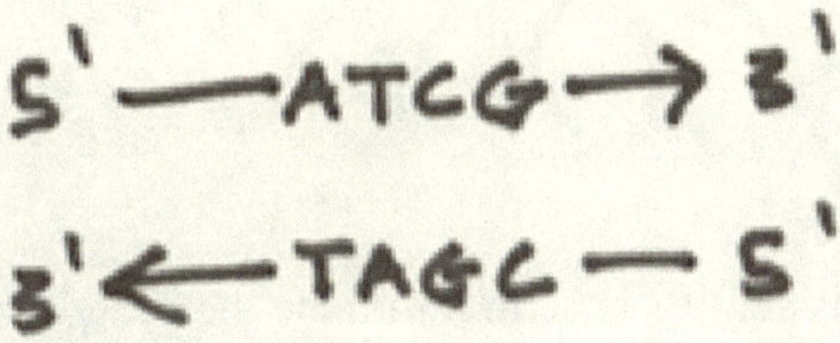

I should reemphasis that the 3' and 5' ends are very different from each other. To be more specific, we say that they are chemically distinct from each other. They are as different from each other as apples and oranges. In fact the 3' end is composed of a *hydroxide group* and the 5' end is composed of something known as a *phosphate group*. These groups look a little like this:

Hopefully, it's easy to see that they are indeed distinct from each other — even more so than apples and oranges. The hydroxide group being comparatively small and meek, whereas the phosphate group is prominent, overbearing even. This turns out to be a crucial factor because replication is carried out by the activities of a variety of different enzymes which all function by focusing on one DNA end or another or both.
So now, the picture looks like this:

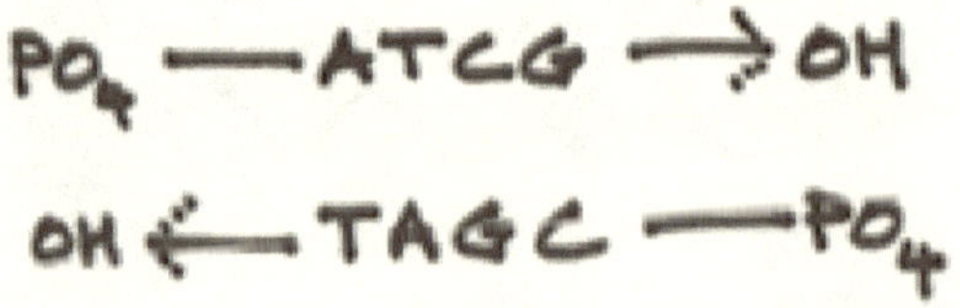

It should also be pointed out that DNA is not really like this flat goofy looking cartoon. As mentioned in a previous chapter, the two DNA strands are actually intertwined around each other in a rather pretty helical fashion. This is where the two strands are wound around each other, sort of like two elastic strings twisted and coiled together. Sort of like this:

Now that the stage is set, it's time to introduce the proteins or the enzymes, which are responsible for the actual process of replication. Enzyme is just a fancy word for a protein that is able to facilitate a chemical process. What

I'll do here is to focus on terminology associated with a simple organism like the bacteria, *e. coli*. However, all organisms, even those as complicated as humans, do more or less the same thing when it comes to doubling their DNA — the principle difference being that unfortunately, the enzymes have difference names and labels.

That aside, the first enzyme for replication in e. coli that we should introduce is, of course, the most important enzyme in the entire process. In e. coli, this enzyme is called *DNA polymerase III*(or DNA pol III for short), and is essentially the one that is responsible for the actual business of making more DNA. If this entire exercise was analogous to a movie, then this enzyme is the marquee player. It is the Tom Cruise, the Julia Roberts, the proverbial bread and butter of replication. It is, quite simply, the star of the entire process. Instead of drawing a picture of Tom Cruise or a picture of Julia Roberts, I think a picture like this should suffice:

Problem is, if we were to draw this enzyme to scale with a helical DNA molecule (like this),

you'll notice that the DNA pol III is actually too big to get inside the DNA strands. It can't go about its business of copying the DNA, because the strands are all coiled up in the helical structure. In other words, there is a serious issue of accessibility. Even our star enzyme, despite its importance, can't do its job without access to the molecules of DNA it wants to copy. Consequently, the enzyme that inevitably has to act first is one that is responsible for opening up the DNA strand. This enzyme is known as a *helicase*, and its role is to essentially unwind the DNA molecule, which would look like this:

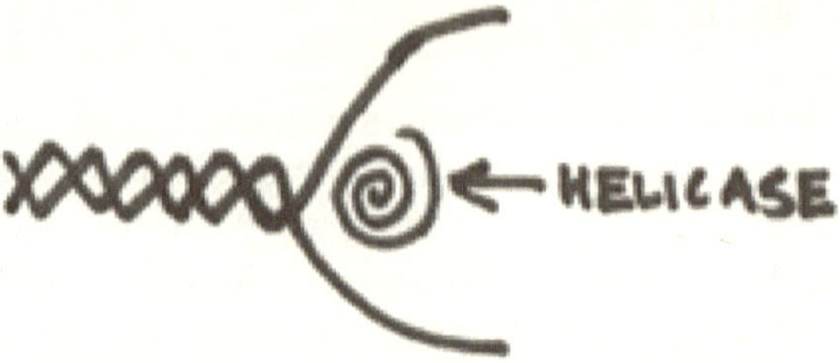

The net effect being the production of a "bubble" of opening where the two DNA strands are pried apart and are subsequently accessible to the whims of the replication machinery.

Curiously, the DNA pol III, which after the unwinding event, can now interact with the DNA molecules, does so whilst attached to a bunch of other enzymes. This attachment is a little like a bunch of buddies hanging out together. The complex actually looks a little like this:

You'll notice it has the following… (i) two DNA polymerase III's: which kind of makes sense given the fact that there are two strands of DNA that need to be copied; (ii) one helicase molecule: which also sort of makes sense, because as this replication complex is doing its thing along the DNA molecule, wouldn't it be handy to have the built-in ability of opening up the DNA molecule as it moves along; and (iii) one new enzyme which is known in *e. coli* as the *primase*. However, the purpose of the primase molecule is a little complicated and so to fully comprehend the role of this enzyme, we need to switch gears a little and tell you a bit more about the DNA pol III molecule.

What actually needs to be done, is for us to go over a few mechanisms that all DNA polymerases seem to use. In fact, it's apparent that every DNA polymerase that has been discovered on this planet:

In fact they all (without exception) seem to follow a two basic rules. Rule number one states that all DNA polymerases function by adding nucleotides to the 3' end of the DNA strand. What this means exactly is that a DNA strand can be extended by the addition of new A's, T's, C's or G's. However, the new nucleotides can only be added to one particular end, namely the 3' hydroxide group. This is a molecular restraint in that the DNA polymerase can only join nucleotides via this smallish chemical group. This rule can be drawn out like this:

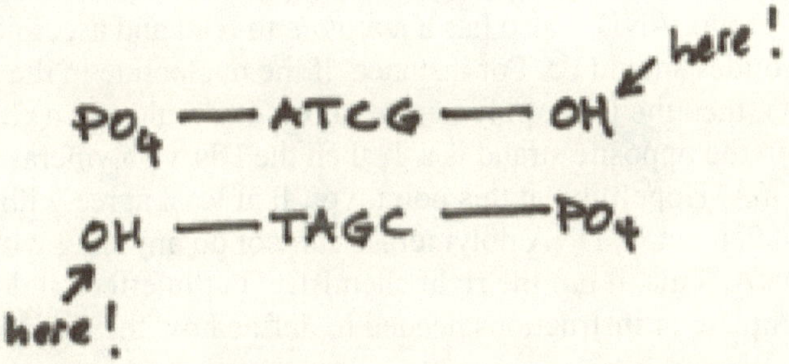

Rule number two states that all DNA polymerases require a *primer* to function properly. This is probably the most challenging concept that needs to be addressed. If you get through this, then you consider yourself home free.

To simplify the notion of a primer, let's look at a single strand of DNA, complete with its 5' and 3' ends. It should look a bit like this:

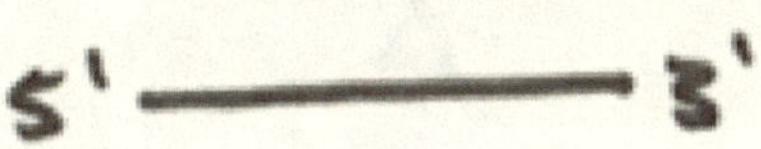

Now according to rule number one, a DNA polymerase can extend this single strand chain but only by adding nucleotides to the 3' end. In effect, you can argue that all of the relevant chemical groups are present for making more DNA. However, the problem lies in the fact that under these circumstances, the DNA polymerase doesn't actually know what to add. How does it know, whether to add an A, a T, a C or a G? It can't exactly be a random event, because replication is all about making sure cells receives an identical copy of the DNA code.

Take the following picture:

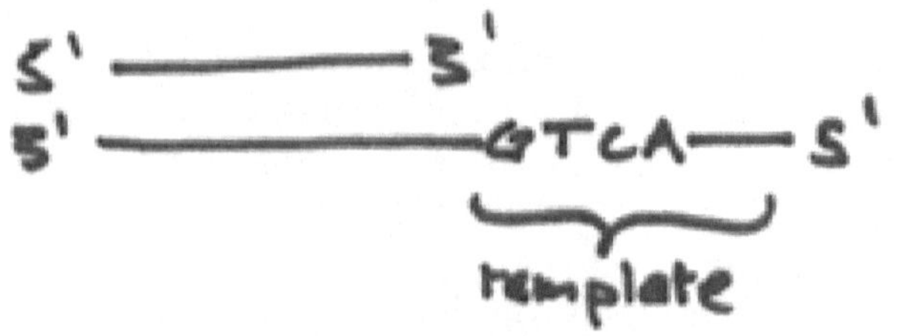

Under this layout, it should be clear that now, the DNA polymerase has the required 3' group, AND it also has a *template* to read and ascertain what those nucleotides should be. For instance, if the nucleotide in the opposite strand is a G, then the DNA polymerase knows it should add a C. If the nucleotide in the opposite strand is a T, then thc DNA polymerase knows it should add a A. Hopefully, at this point, you'll at least agree with the following statement. A DNA polymerase can not do anything with a single strand of DNA. True, it has the right chemistry, but in effect, it does not have the template or instructions needed to define how the chain is extended.

If we redraw the picture. Say like this:

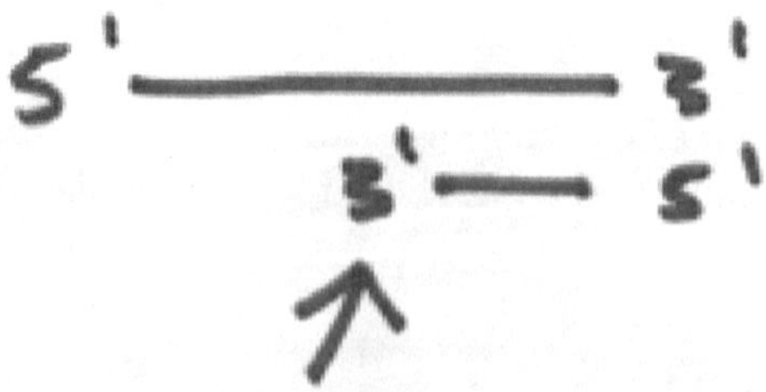

What you'll notice are two strands of DNA, one long and one short. You'll also notice that the strands are anti-parallel as discussed earlier. If you focus on the arrowhead, you'll find yourself focusing on a perfectly situated 3' group. Here is the end of a DNA strand that is chemically ready to have nucleotides attached. Furthermore, it is also a 3' end that is located where a template is present on the opposite strand. In other words, everything is in place. The right chemistry, and a means for instructing which nucleotides to add. Again, taken at the simplest level, we can conclude that in order for a DNA polymerase to do its thing, it needs an area of double strandedness. So,.. the small sequence of nucleotides that has been circled here…

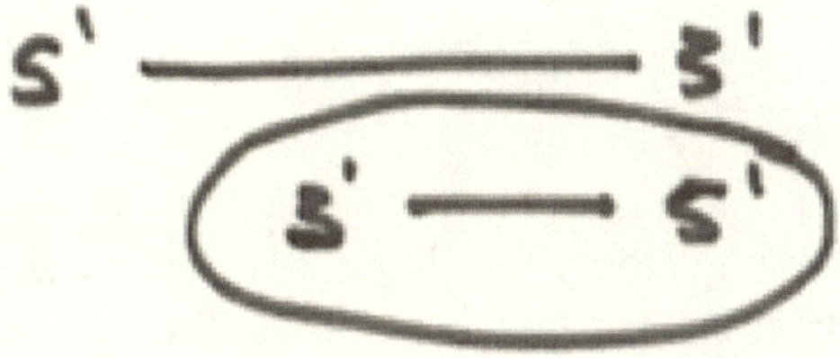

… which makes an area of double strandedness is technically known as a primer. With this all sorted out, hopefully the rule about requiring this primer makes a little more sense, and you can probably guess that the enzyme called the primase may have something to do with this nuance.

Which turns out to be exactly what this primase enzyme is all about. In a nutshell, it is an enzyme capable of making a short sequence of nucleic acids which functions as a primer. A key point that needs to be emphasized, however, is that this primer is made up of RNA, which if you recall, is a molecule that is very similar to DNA in that it is also composed of the representative four nucleotide code. This is actually due to a biological technicality whereby it is possible to make a complementary strand of RNA without the use of a primer (hmmm, think about this for a second). Taken together, the function of the primase should end up looking like this:

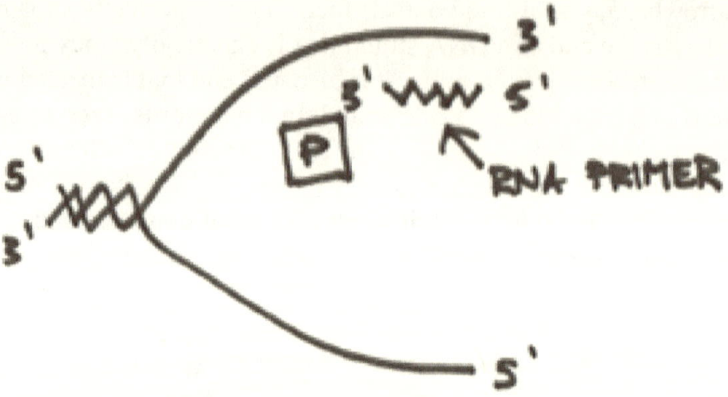

If you've been following along, then hopefully you can see that replication from this RNA primer can proceed in a manner that can be drawn like this:

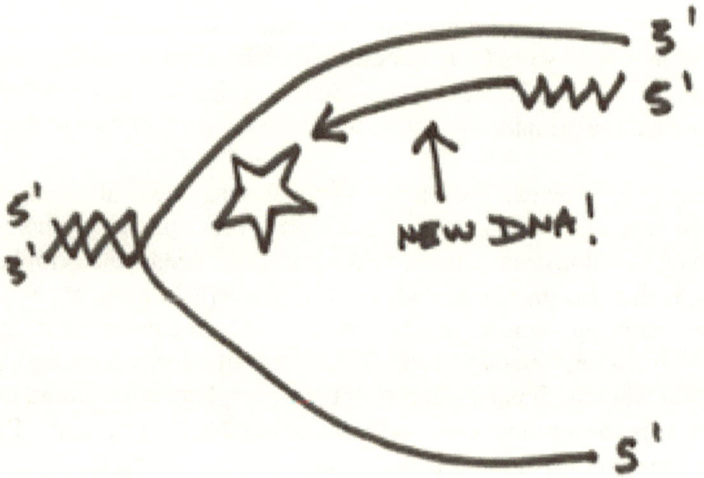

However, it's wise to pause here for a second, because you have to understand that whilst this top strand is being replicated, the lower strand is also being worked on simultaneously (There are two DNA polymerase III's attached together afterall). The lower strand is actually a bit messier for reasons that will become clearer as we proceed in this discussion.

Basically, the primase enzyme will also go about preparing a primer for the lower strands. However, if we draw this primer and label the ends in the anti-parallel manner, you can hopefully see a logistical problem in this set-up. Take a look at the following picture, and see if you can find the problem (remember, the DNA polymerases, the helicase and the primase all move as a single unit in one direction, and remember that all DNA polymerases must add to the 3’ end):

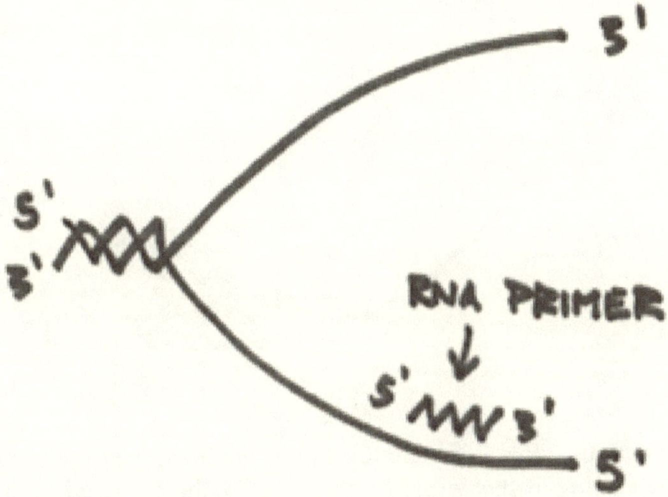

Do you see the problem? Do you see a problem with the direction of the primer? Do you see that the 3’ end of the lower primer is facing the wrong direction?

This is obviously a problem, and it turns out that in order to overcome it, the DNA polymerase will still add nucleotides to the 3’ end, but can only do so for a short distance. To keep it simple, think of it as being able to replicate as far as the enzyme is big, which should look a little like this:

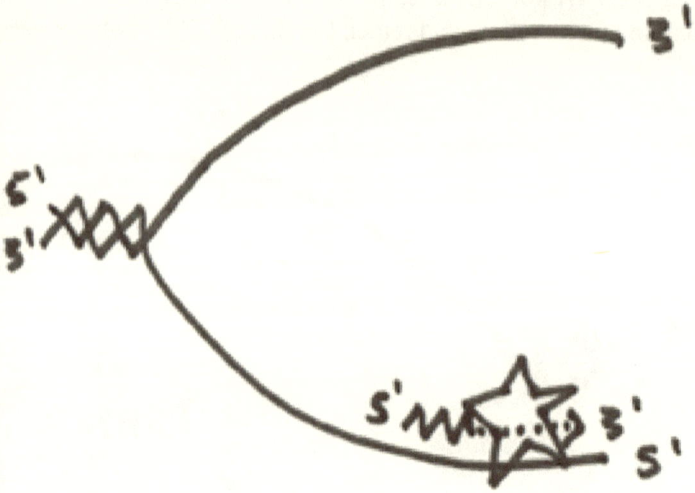

Unfortunately, this doesn't inherently solve the direction problem, so what ends up happening, is that with this lower strand, the primase has to continually make a primer, and the DNA polymerase III has to continually replicate a little bit. In the end, it should look like this:

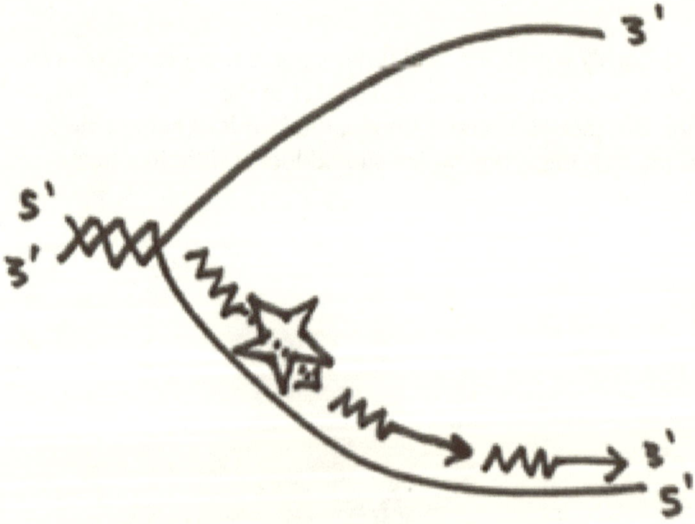

The difference in how each strand gets copied is reflected in why some people call them the*leading* and *lagging* strands of replication. One strand is obviously fairly straight forward whereas the other is quite labour intensive.

Anyhow, after this is all said and done, hopefully, you'll agree with the following statement. That is, we have finally doubled or copied our genetic sequence. However, it should also be clear that the whole thing is a bit messy. For instance, there are bits of RNA everywhere, and the lagging strand is composed of pieces. To address these problems, we have to introduce a few more enzymes.

The first of which is *DNA polymerase I*, which I will draw as a fish with sharp teeth. This enzyme is special in that, in a nutshell, it is responsible for dealing with the RNA. In a nutshell, its job is to somehow replace it with DNA. In a nutshell, I'll draw it like this:

DNA polymerase actually has two distinct functions. Firstly, as its name implies, it is a DNA polymerase, meaning that it is capable of extending the DNA chain, but in doing so must follow the same two rules that govern these enzymes. In other words, it must add nucleotides to the 3' end and it must use a primer as a springboard. Ironically, it is a shitty DNA polymerase. Whereas DNA polymerase III can replicate for several hundred nucleotides, DNA polymerase I has difficulty getting past a few dozen.

Secondly, DNA polymerase I is also an*exonuclease*. This means it's capable of degrading or chewing up nucleotides. Which is another reason why I drew a fish with teeth. And not only does it chew stuff up, it does so in a fairly specific manner. To begin with, it likes to start at areas, which are termed as nicks in the DNA. In our picture, this is where the nicks would be:

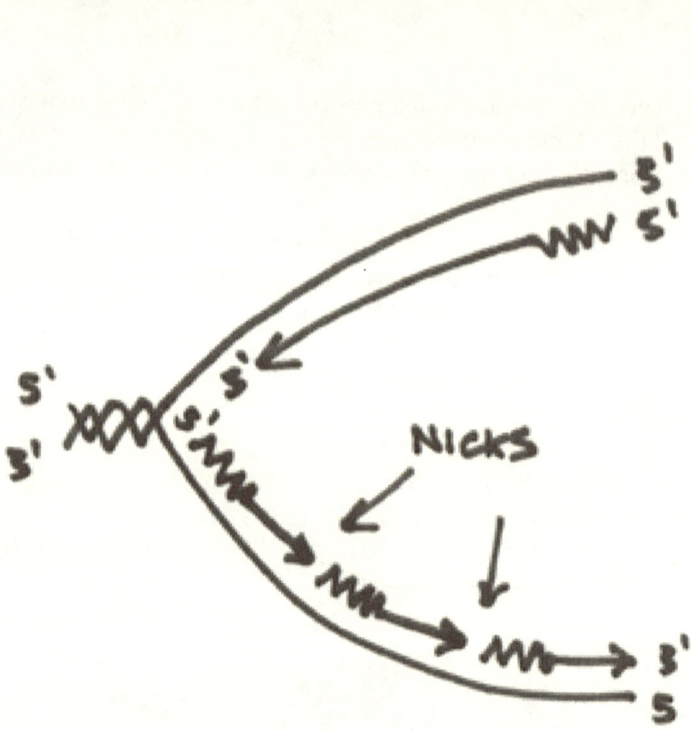

Furthermore, this exonuclease is picky in that it always chews from the 5' end. Basically it is gunning for that big phosphate group. So that you don't forget this, I've drawn this picture to help you visualize this:

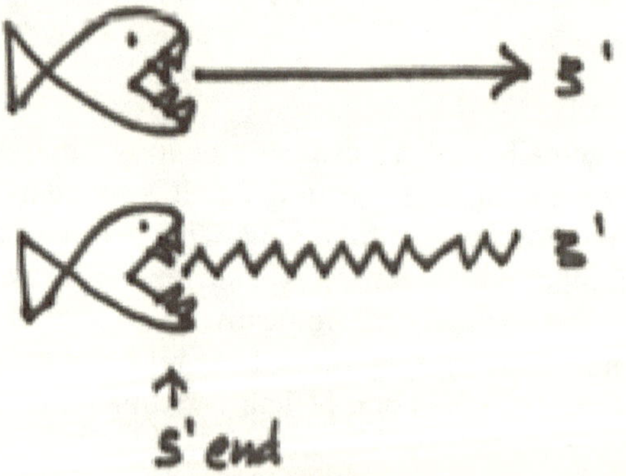

Now, if you take all of this into consideration, you come up with the following mechanism. DNA polymerase I will come in on our replication picture, and zone in on a nick in the strands. Once there, it will begin chewing on the 5' end, which should look a bit like this:

Don't forget that this enzyme is also a DNA polymerase, and if you look at the other side of the nick, you will hopefully realize that there is this beautiful 3' end ready for action. This beautiful 3' end is right here:

Let's say that the fish's ass happens to contain the DNA polymerase function. What therefore happens is that DNA polymerase I will start replicating from that 3' end, which incidentally fills up the gap that was created by the exonuclease activity. This should nicely demonstrate how DNA pol I achieves its function of replacing the RNA with DNA. This whole step should kind of look like this:

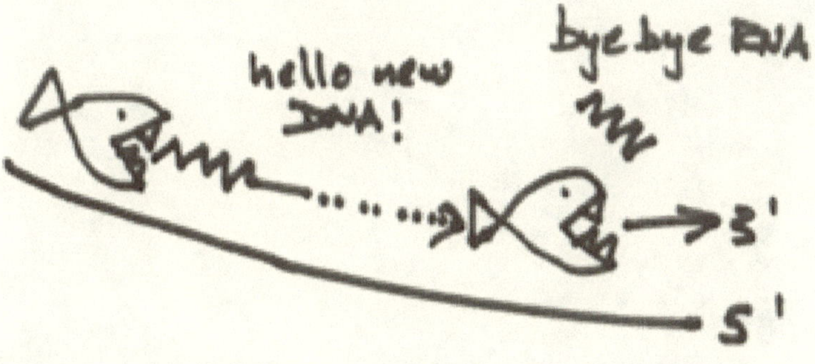

Hopefully, this puts the shittiness of this DNA pol I in perspective. It's quite biologically pretty because, I hope you can appreciate that DNA pol I doesn't need to be very good. It's only responsible for replicating the small region encompassed by that RNA primer.

So,.. after this enzyme has done its thing, you should now agree with the following statement — that you have now doubled your DNA. Of course, it's still a bit untidy because the strands (especially the lagging strand) are still in bits and pieces. Enter the next and final enzyme, which is called the *ligase*. This enzyme has only one job and that is to seal all of the bits and pieces together. It fairly analogous to a glue job and essentially your picture will go from something like this:

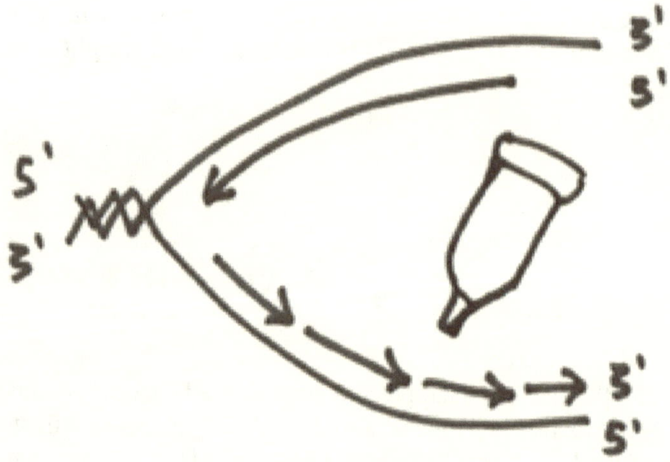

To something like this:

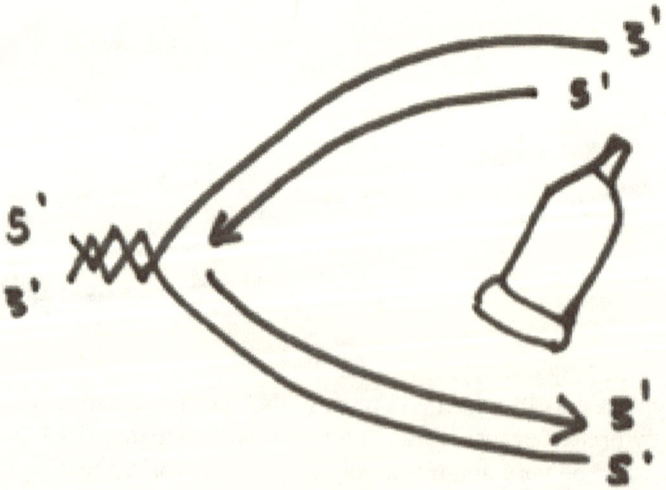

And (drum roll please) VIOLA! You have doubled your DNA. You have made two copies of the same genetic code – which during the process of cell division, will enable each of the two new cells to receive a copy of the genome.

One of the nuances that should be mentioned is that if you examine the entire process, you will notice that each of the DNA sequences is derived from one old strand and one newly synthesized strand. Because of this, replication is often termed *semi-conservative*, whereby each of the original two strands is read individually to synthesize a new and complementary strand.

* * *

Actually, I lied. It's not quite over. Before, I finally put this whole replication thing to rest, I think it's also worth talking about one other enzyme, or a family of enzymes, known to scientists as*topoisomerases*. I like mentioning these enzymes, because I think they do a wonderful job of illustrating just how complicated and elegant nature is, when confronted with a specific job.

What we'll need to do here is undergo a visual exercise. Let's say I tell you to hold two fingers up like this:

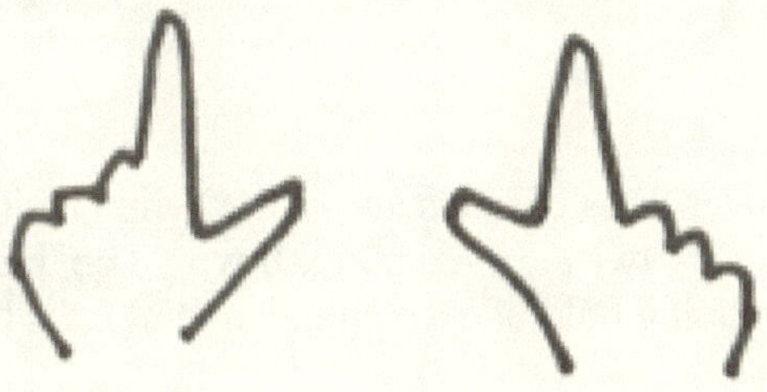

And let's say that I have an elastic band. With this elastic band, I will twist and coil it and then place it around both of your fingers. Essentially, this will represent the double helix and will look a bit like this:

If you recall, the first thing that had to happen was for a helicase enzyme to come in and open up that helix structure. Let's say that I am the helicase, and I come in and grab hold of the two strands of your elastic band and pry them open. It should make a little bubble and should look a little like this:

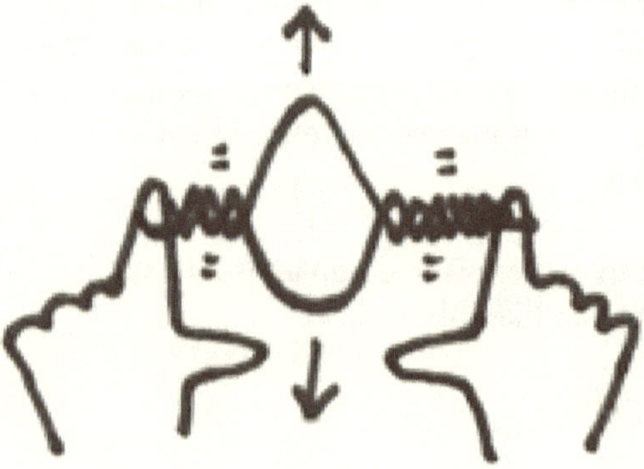

Can you see that under these circumstances, the helix on either side of the opening will be actually twisted even more. It would be like taking your replication fork, grabbing hold of each strand, and like the helicase forcing an opening like this:

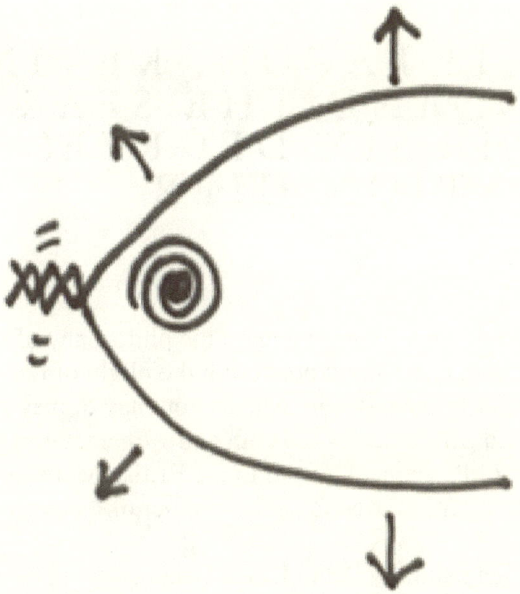

Do you see that this will cause a further tightening of the coil along the helix?

This is actually very bad for the DNA molecule, as this twisting can cause a lot of structural stress. So much so, that the DNA molecule is in very real danger of snapping – which you can imagine would be a very bad thing to happen during replication.

Topoisomerases are enzymes that are designed to take care of this problem. These enzymes can actually detect these areas of high structural stress, and zone in on them. Not only that, but whilst they are at these areas, they will then cut both strands in the DNA complex. Remarkably, they will then hold on to all four ends of the cut, and in a very controlled fashion, unwind to alleviate the stress. Finally, they will also behave like ligases and stick back the correct ends together again.

This is nothing short of amazing, and hopefully you can see that these enzymes play an important role. As the DNA is opening up for replication, there will always be an issue of structural stress, which is always addressed by the actions of these remarkable enzymes.

Anyway… taken together, that, Mr. Trout is what replication is all about.

FRANCIS BACON, KEVIN BACON, AND THE SEARCH FOR THE SIX DEGREES OF SEPARATION HIER

Lately, I've been doing a little writing on the philosophy of science, and a consequence of this, is my mind pondering the plight of Bacon. Not the food, but rather Sir Francis Bacon, who as you may or may not know, is the renown writer and gentlemen of the 16th and 17th centuries – famous for being a member of Parliament, friend to the British Monarchy, and (most important to me) often referred to as the "Father of the Scientific Method."

Such thinking then naturally led to Kevin Bacon, who in turn, reminded me of the "Six Degrees of Kevin Bacon." Which is also to say that inevitably, I landed at entertaining the specifics of the "*Six Degrees* of Sir Francis Bacon."

This refers to the phrase, "The Six Degrees of Separation," which submits that you are less than six "friend of a friend" steps away from everyone else on the planet. In other words, it suggests that mankind is more connected than you would think. Interestingly, this calculation has never been formally proven, and there might even be some evidence to suggest that social media has brought it down to four degrees, but despite all this technical wrangling, it is nevertheless obvious that it probably only works well if the people involved happen to be *alive*.

Which is to say that the "Six Degrees of Sir Francis Bacon," a man who died in 1626, are probably all dead.

With this in mind, we need to return to the "Six Degrees of Kevin Bacon." Whilst, this originally concerned itself with connections in the entertainment industry, the phrase nowadays is kind of symbolic of humanity's interconnectedness. Put another way, Kevin Bacon is a little like an unofficial figurehead of this game.

But figureheads are usually transient. Indeed, the fact of the matter is that Mr. Bacon is no longer the sprightly young man that danced into our hearts in *Footloose*. Nor is he, despite the fact that he played an "invisible" character in *Hollow Man*, capable of hiding from the debilitating onward march of time. In essence, he should be fully aware that as he ages, the concept and the mathematics of the "Six Degrees of Kevin Bacon," will no longer be practical – indeed, it will no longer even be relevant. Consequently, you might suppose that one day, there will need to be a proper discussion about a "six degrees" successor.

And why not start that discussion now? Namely, how would one decide on such a successor? Of course, this would come with a few rules. For instance, whoever is chosen should, at the very least, be younger than Kevin Bacon. Perhaps Mr. Kevin Bacon should even have a role in this process. Anyway, as I continue to procrastinate from doing whatever it is I am supposed to be doing, I'd like to put forth the following scenarios and then maybe see if the procrastinating community at large has any thoughts on the matter:

—

1. The British Monarchy model.

This is where the weight of responsibility is passed on to the first born. Furthermore, since we're being thematic and all, this option should totally include a throne and also a crown that can be worn on special occasions. Maybe a fancy sword as well. Yes, a sword would be totally awesome – "The six degrees of so and so and his/her sword" has a nice ring to it.

—

2. The Democratic Model

Why not do this with an open election? This would certainly be entertaining to watch, and would no doubt fuel some interesting discussion. Although the mind boggles at how the nominees will be decided upon, and how exactly they would present themselves (more so, since the principle of the Six Degrees, hypothetically is meant to be immune from the nuisance of ideology).

—

3. The "So You Think You Can Dance" model.

This would be the obligatory "how can we turn this into reality TV" option. Furthermore, as Mr. Bacon, himself, is no stranger to the entertainment industry, it is perhaps the most logical model to find a successor. A dance off, moreover, would be nothing less than magical. Think of the how fun this might be, think of the spectacle, think of the press, and think of the Kevin Bacon themed So You Think You Can Dance stationary. As well, each time a successor is chosen, the theme of the next reality show could be tweaked according to the accomplishments of the new figurehead. Imagine different contests each time around, ranging from cooking to planning a wedding, to a full on Hunger Games styled competition.

—

4. The Kevin Bacon as an Eternal Deity Model (and the similar themed "Kevin Bacon Reincarnate Model")

Let's face it – maybe Kevin Bacon would rather keep all the glory to himself, and also keep it forever. If so, there is another option out there. Both Jesus of Nazareth and Kim Jong-il of North Korea used it. Basically, it's where Kevin Bacon declares himself the reference point, and instead of looking for a successor, the actual number of degrees changes with time. In other words, in a few years, we can call it "The seven degrees of Kevin Bacon," and then "The eight degrees…" and so on and so on. Alternatively, it could be like the Dalai Lama, and every time you pass on, there is a reincarnated version of you being born elsewhere. I am not sure how this would work exactly (how would we identify this reincarnated Kevin Bacon?), but it seems to me a reasonable idea. Plus, the thought of an organized religion with the word "bacon" in it has great appeal.

—

Anyway, it would be interesting to hear of any other ideas, or even better, to hear a successor suggestion or two. As well, let me just end by saying that if this all sounds a little too complicated, then let's simplify things and just pick me. I would totally be down with being next in line – especially if I can somehow score a throne, crown and a sword out of the deal.

IN CANADA, AFTER ANY INTERNATIONAL CLIMATE CHANGE CONFERENCE: I FEAR CORRESPONDENCE OF THIS SORT WILL BE SENT

"In stark contrast to its cuddly international image, Canada is the dirty old man of the climate world – missing its Kyoto emissions reduction target by a country mile (by 2007, it was 34% above its target) and showing no signs of reigning in its profligacy."
The Guardian, November 30, 2009

>
Petey,

What the hell is going on? That conference was a freaking fiasco! What happened? And how is Mr. Environment Minister going to do to fix it?

Stevie (The PM).

>
Steve buddy!

O.K. We have a plan. A couple of things actually. Most of them revolving around science and stuff, since we keep getting hammered on our stance with what the climatologists are telling us (you know, the IPCC reports and such). Anyway, the plan is multifaceted, and we're still bouncing off ideas (FYI: if you got any Prime Ministery input, just pass it on), but here is what we have so far:

1. To get the scientific community off our back, we're going to challenge them to perform *definitive, but basically impossible*, climate science experiments. Doesn't that sound great? I wrote that myself. And here's one just off the top of my head, which I'm calling the TRI-EARTH experiment (also, wrote that myself). Here, we'll ask scientists to create two other

planet Earths, and populate them with identical geology, biodiversity and anthropogenic infrastructure, and then do a compare and comparison. Our current Earth could be the test subject, whereas the other two could represent "controls" (ooh actual science lingo). These would be conditions with (a) zero fossil fuel emissions, and (b) intensive fossil fuel emissions. Scientists would then be asked to collect data for 100 years, and then reconvene with their conclusions. Brilliant right? Oh man, our tech guys are gonna love making that website.

2. To get the environmental community off our backs, we're thinking of asking the HR Departments of all tar sand companies to actively hire members of the biodiversity community. And we're not talking scientists here, but *actual* animals – the cuter and the furrier the better! Anyway, the idea is that this would be an excellent way to create tension between all those environmentalists. Imagine the debates! I can hear them already: "*You can't shut down the tar sands! Think of the livelihood of our friends, the [insert name of cute furry mammal]. How will they maintain their way of life?*" Basically, with the right amount of nuts, we could get a squirrel or two to say anything. As an added bonus, the irony alone just might get Suzuki's brain to explode.

3. This one is a biggie! We're looking into actually creating new scientific laws! Wouldn't that be great? I mean a good chunk of the data out there is based on rigorous climate modeling, which is powered by scientific laws and mathematical equations (bla bla bla). So we say: why not take matters into our own hands, and create something like a new addition to the Laws of Thermodynamics. I mean, these laws are well known, they come up a lot in climate studies (the first law with its overbearing "energy cannot be created or destroyed" mantra is especially annoying), and as a bonus, they even have too many syllables which we know is always good for added confusion. If we're smart, we can even make the new law a little "magical" (seriously, maybe something about unicorns – you like unicorns right?). This might make the whole creationism angle a little easier to swallow scientifically (and you know me, I'm always looking for ways to widen our support base).

4. Advertising: and lots of it. Maybe go with either a "Canada is a Climate Change Free Zone" angle (wouldn't that look great on a t-shirt?); or maybe just a straight up promotion of things to do in a hotter climate. I think the "Hot Canada" idea could sell itself. I'm thinking five words: beach volleyball and umbrella drinks. Hmmm… let me write that down. Could work as a possible slogan.

O.K enough writing… I'm going to send this memo off right now. These are just a few ideas we're ready to act on. Add on a good old general marketing blitz, and I think we got something that should do the trick. Anyway, just say the word boss and we'll get on it pronto.

Petey

>
Petey,

Sounds great. Make it so (I love saying that). Oh and how about this for a slogan, "No more sweater vests!"

Later,
Stevie.

HAN SOLO AND CHEWBACCA WEIGH IN ON THEIR NEW HYBRID MILLENNIUM FALCON

HAN SOLO: Well, so far, it seems like it's a pretty good thing. Me, I'm not too up on the technology, but Chewie is pretty good at that stuff. Right Chewie?

CHEWBACCA: Uuuhhhggg. Rrrrggghhh. Hhhgg-aaa. Rrrrn.

HAN SOLO: Yeah, that's a good point. Chewie just reminded me that this new system has significantly increased our energy efficiency. This basically means less money spent at the pump, and more money in our pockets.

CHEWBACCA: Rrrrrr! Aaaa-Ghhhuuurr. Uuuuhggg.

HAN SOLO: Right. And lower emissions too. Although I don't get why that would be such a big deal in deep space. Do greenhouse gases do anything out there anyway?

CHEWBACCA: Uuuuhhh-rrrr. Ghhhgggg. Uuugggg. Ggg. Rrrrr-uuuuaa. RRRR! NNHHHUUUR!

HAN SOLO: Alright, alright. Calm down. I'm not saying it's not a problem. I know there's science behind all this stuff. It's not like you haven't told me to be environmentally conscious like a hundred times already. Look, I'm sorry buddy. I didn't mean to sound negative like those Empire bastards.

CHEWBACCA: RRRR! RRRRRRRR!

HAN SOLO: Yeah, I know. That *would* be pretty funny to watch you pull the arms off a one of those guys. Doing that would be carbon neutral too right?

CHEWBACCA: Gghhnn. Nnnnh.

HAN SOLO: Yeah, sure. But listen Chewie, seriously: How would lower emissions in deep space help? I just don't get it, you know?

CHEWBACCA: Grrrrgh. Uuurhh. RRRggllhh. Hhuu-hhhuu. Auhhh-ghu-gh. RRRRR!. Ggg-rrr, uurrghh. HHGGU! Uuuuhh. Rrr, ggghhu. Huuhhhg. GGGrrr. Uhh?

HAN SOLO: Oh, O.K.. That makes sense. You say you still want fewer emissions because there's still a lot of flying involved when the Falcon leaves or returns to a planet, or just when she does her cool maneuvers close to the surface. These things still directly contribute to increasing greenhouse gas amounts within the confines of the planet's atmosphere. Hence, not helping with the global warming problem.

CHEWBACCA: Ggggrrr. Rrrrh. Uuuhhggr. RRRR! Uhhfuckinggghug.

HAN SOLO: Definitely. And you're right, Tatooine is already too damn hot.

CHEWBACCA: Rrrrrhhg. RRRGGH! Hhhuurrg. Ggrrgh. Huurg. Grrhhg. Guuuaaauu. AAAURRGG! RRRRGGG!

HAN SOLO: Yeah, O.K. I mean I'm basically pretty happy with the modifications. Really, as long as we can still make the Kessel Run in less than twelve parsecs, I really don't care. Plus, I still get to say stuff like "Punch it Chewie," right? Chewie, you love that stuff.

CHEWBACCA: Ggrrrrgghhaarr.

DNA AS A MAGIC 8 BALL: CONCERNING THE PRESIDENT OF THE UNITED STATES

Written in 2006, this piece attempts to do character assessments of US political figures based on DNA similarity. It was freaky, truly freaky actually.

Every living thing on this planet adheres to a script, a biological language that is not unlike the ingredient lists on the back of your grocery store products. This script is DNA, composed of a limited alphabet of four building blocks (or letters, if you will): A, T, C and G.

Our human document is just over three-billion letters in length. To offer some perspective, *E. coli* has just over four-and-a-half million letters, a fly has about 150-million letters and rice has close to 400-million. In all—as of August, 2005—over 100,000,000,000 letters of code have been sequenced from a multitude of earthly delights and made publicly available for research within the life sciences.

DNA orchestrates the production of proteins—the molecules that are responsible for the architecture, mechanics, senses and defenses of each and every cell and tissue in an organism's being. These proteins actually do the work of "living."

And here's where it gets interesting: Proteins are composed of strings of amino acids, pieced together as a direct result of DNA code. There are 20 different amino acids, each one denoted by a single letter. Since amino-acid alphabet is only missing the letters B, J, O, U, X and Z, one can look for relevant words within the huge dataset of genomes—within life's code—and, perhaps, find wisdom for important decisions.

With this in mind, I decided to supercollide genetics and politics—more specifically, to contemplate specific words, built with strings of amino acids, and search all available genetic and protein sequence data for relevant matches[27]. And it is these matches or answers that are gleaned—as if from a Magic 8 Ball—to reflect and evaluate our leaders, our options and our future. Whether you buy into this brand of decision making or not, here is what you'll discover when you search genetic code for amino-acid sequence strings such as "BUSH," as well as other names from current events.

1. The query for "BUSH" receives no hits, primarily because it is deemed a "low complexity sequence." This is compounded by the fact that the letters B and U do not exist as specific amino acids.

2. To be fair, I tried the string "GWBUSH." Here, the closest match resulted in the sequence "GWDASH." It was interesting to note that 21 of the top 22 matches were derived from the genomes of "uncultured" organisms—ones that cannot be grown in any laboratory setting

3. Next, I tried "GWBLISH," under the pretense that when you squint, it looks like "GWBUSH." In this case, the best sequence match referred to the Japanese strain of *Oryza sativa* (paddy rice), a food staple from a country that is justifiably sensitive to past actions of the United States.

4. Because none of the above results sounded particularly encouraging, I figured that a better indicator of Bush's worth might come from querying the names of his top advisors. However, when the sequence strings "ROVE," "RICE" and "ALITO" are queried, all are met with the "low complexity sequence" result. The top hit for "RUMSFELD" was *Xylella fastidiosa*, a grapevine-decimating pathogen infamous in the wine industry. Interestingly, the top two matches for "CHENEY" are *Vibrio vulnificus*, a bacterium in the same family as those that cause cholera, as well as *Vibrio speldidus*, a nasty intestinal pathogen known for inducing vomiting, diarrhea and abdominal pain.

5. Finally, in an effort to further demonstrate my impartiality, I begrudgingly entered "PRESIDENTBUSH." In this case, the best non-hypothetical match—one that can actually be assigned a biological function—was from the genome of *Entamoeba histolytica*. The organism is a single-celled, parasitic protozoan known for infections that sometimes last for years, which may be accompanied by vague gastrointestinal distress or dysentery—complete with blood and mucus in the stool.

---

[27] Anyone can do this with a common bioinformatics tool known as BLAST. Follow the link and click on the "search for short, nearly exact matches" under the PROTEIN subheading. In the new page, enter your query, and then hit the "BLAST" button.

6. For good measure, I considered how 2005 could have been different politically, entering a couple of searches related to Senator John Kerry. A query for "KERRY" received many perfect hits from a wide variety of different organisms, and "PRESIDENTKERRY" results in a best non-hypothetical match to a gene in *Zygosaccharomyces rouxii*, an organism belonging to the kingdom fungi.

I'm left to make the following conclusions: Simply stated, Bush is of low complexity. Addressing Bush using his first and middle initials suggests that he will run away or act in an uncultured manner. Squint at Bush and he just might make fun of your slanting eyes, call you "sushi lover," or make some other inappropriate comment. His closest advisors are, at best, too simple for the task or busy attacking wine, or, at worst, will make you suffer horribly. At the end of the day, if you don't want the hypothetical, but instead want the truth: President Bush is akin to an extended period of significant discomfort in your gut.

In hindsight, it would appear that these queries reflect accurately on the past year, what with the general mismanagement of the Iraq war, the fallout from Hurricane Katrina, the administration's rebuff of climate change, as well as the President's awkward but accommodating tone with intelligent design.

But, what if John Kerry had been elected president? Well, he would have just been a "fun guy!"

Perhaps there is some merit to this method of divination after all?

As a postscript—and to look forward rather than back, it being a new year—I ran one last query. Given Hillary Clinton's potential candidacy in the 2008 presidential race, I performed one last search, inputting "HILLARY."

Here, the top non-hypothetical hit corresponded to *Burkholderia vietnamiensis* strain G4, a bacterium known for its prowess in eliminating various hazardous environmental contaminants that are found in groundwater. Could this result possibly foreshadow an interesting campaign ahead? Is HILLARY someone that will bioremediate—literally "clean up" —the polluting mess left by the current administration.

No matter, I think it is best that I leave that act of interpretation to you the reader, since my actual Magic 8 Ball suggests that I "better not tell you now."

A BIOLOGIST IN NIGERIA

Dr. Oyekanmi Nashiri is a busy individual who exudes enthusiasm, embraces optimism, and covets high expectations. Then again, as the principle organizer of a somewhat curious scientific program, he would have to be all that and more – some would even say that his good intentions place him squarely in the category of certified nutbar. Nash (as he prefers to be called) has spent the better part of his scientific career developing and implementing the West African Biotechnology Workshops, a focused attempt on bringing scientific expertise and potential research collaborations to his homeland, Nigeria. Which is to say, he is intent on bringing the realm of high technology into an otherwise struggling country.

Ironically, I was thrown into this mix by virtue of my reputation as a university instructor, the lure of traveling to an altogether foreign place, and the somewhat naïve notion that this challenge could bring some merit into the developing world. Ironic because part of the appeal stemmed from my being guilty of harboring the same preconceptions about Africa that every other non-African seems to have. That is, the whole romantic 'Out of Africa' thing, where the nations that hold the continent together live in natural but primitive splendor.

"Which," as Nash would often say "is all nonsense." Nash is an animated speaker – his continuous gestures and movements betray his scientific patience. "You can't think of Africa as one place, one culture. It is distinction within distinction. Every place is separate and special from the other. We are not going to Africa, David, we are going to Nigeria." And in retrospect, he couldn't have been more correct.

Nigeria is a country of unfathomable extremes, the kind that Meryl Streep and Robert Redford would take little comfort in. It has a population of over 100 million individuals crammed into a small wedge of land that is the coastal armpit of West Africa. Its growth rate is such that the population is expected to rise to about 300 million individuals by 2025 – a figure that would mirror that of the United States, except that it would be squished into an area half the size of Alaska. Such numbers also give rise to extreme cultural diversity, which is well exemplified by the more than 250 different ethnic groups, of which three stand out and comprise 65% of the population

(Hausa in the Muslim north, Igbo in the agrcultural lands of the south-east, and Yoruba in the urban south-west.). In spite of this dominance, the remaining groups certainly do not want to be ignored. Each has its own language; each has its own way of life; and each has its own fiery brand of pride. In short, Nigeria is as close as it gets to a real cultural melting pot, African or otherwise. Quite simply, it is distinction within distinction within distinction.

It is also not a very nice place to visit. The travel advisories within the Canadian Department of Foreign Affairs and International Trade say it all rather succinctly:

The official language is English. Tourist facilities are limited. Power shortages and low water pressure are common. Telecommunications are unreliable. Those attempting to contact the police may have difficulty getting through. Violence and unrest sparked by tensions between ethnic and religious communities occur in various parts of the country and have resulted in numerous deaths. The military may intervene and curfews may be imposed. Canadians should always maintain a high level of security awareness and inquire about local conditions when travelling in the country.

To be more specific, our final destination was actually Lagos, a crowded and polluted city of some 13 million inhabitants, and a city with the dreadful reputation of having one of the highest crime rates in the world. In fact, the lonely planet guide I picked up heartily recommends it for the "truly masochistic voyager." And upon reading through several sources of information, it really seemed like it would be wise to avoid, by all counts, practically everything.

To these information nuggets, Nash would quickly rebuttal. "This is all nonsense. Do not worry about such propaganda. Nigeria is the West African superpower. We have one of the most highly educated workforces in the continent …" And then in the middle of the blur that is his hand gesturing, he would pause and smile, "and we are going to win the World Cup."

Still, as an educator who teaches pupils and the public alike in the nuances of biology, I wasn't necessarily swayed by his arguments. A country's soccer prowess holds little weight in teaching scientists the practical and theoretical aspects of molecular genetics (for instance, I have yet to witness an offside penalty in my lab). But things like reliable power and water sources do matter and, as Kate my wife would so ardently point out, it's also nice to not worry about things like violence and riots.

* * *

The morning I left for Nigeria was a bit of a somber affair in my homestead. I would be away from my wife and my baby daughter for two and a half long weeks, the first time I was to be away from them as a new parent. This wasn't really how it was supposed to unfold. Originally, this teaching assignment was scheduled to be held in a previous year, where Hannah would be but a twinkle in our eyes or at the very least, safe in a second trimester haven – both situations that wouldn't have yet revealed the emotional enormity of fatherhood. It was originally to be an adventure. Even Kate had toyed with the possibility of joining me. However, adventure would now take a secondary role to family life and Kate would join me only in spirit and kind words via a bundle of letters – one for each day of my trip. It seemed that she took to heart the warnings about unreliable telecommunications.

Thankfully, I was not traveling alone. I enlisted the help of Dr. Samantha Greunheid, a post-doctorate and colleague at the University of British Columbia. Sam was a specialist in infectious diseases, which seemed to be a good fit given our destination. She was also knee deep in proteomics, a burgeoning research field that looks at proteins (what the DNA genes code for,) and, for lack of a better description, tries to look at all of them in any given circumstance. For instance, Sam was primarily interested in looking at proteins present in her bacteria, and in particular trying to observe differences between samples that were docile, and samples that were actively involved in an infection. In this respect, proteomics is a newer breed of science that relies heavily on high technology, high throughput and almost steamroller-like tendencies, all the while generating massive amounts of data such that mathematicians and statisticians are being courted into the process. This part of her research would be completely novel to the primitive state of Nigerian science. It was to be a compelling and interesting mix. Most important, however, Sam was down to earth and had an easy low maintenance attitude, which I feared was going to be necessary for this trip. Before we had even departed on our flight, it became clear that our itinerary was a focal point of some intrigue. I remember the conversation we had in Seattle with one of the immigration officers who examined our visas. It began innocently enough:

"Where are you going?"

"Nigeria."

Then came a short pause. "Nigeria eh? Why are you going there? Are you missionaries?"

"No, actually, we're university teachers. We'll be teaching a science course there."

Another pause. "Well,.. better you than me." A sidewise glance, "You know, you may as well say goodbye to your luggage now. Those folks down there, they just take whatever they want."

* * *

How Lagos, and Nigeria in general came into these troubled circumstances is rather neatly if not bluntly explained in the book "The World's Most Dangerous Places," a Christmas gift given to me by my not so subtle parents. In it, Robert Young Pelton – who could give the Crocodile Hunter a run for his money for pure bravado – provides descriptions and anecdotes for destinations you would best avoid. My worried mother had actually folded down the pages on Nigeria, such that its 3 out of 4 'peril rating' was immediately evident.

In short, Nigeria had for the better part of its 40 year history been subjected to the whims and follies of various military rules. Which together with its diverse cultural mix and its lucrative oil deposits, exacerbated into an unwieldy recipe for political chaos, ethnic tension, and rampant crime. However, even the casual follower of world politics knows that what makes Nigeria especially infamous, is its almost legendary propensity for corrupt practices. In fact, Nigeria has consistently ranked as one of the worst offenders in various Corruption Perception Indexes, and it would probably surprise no-one if its unclean borders were shown under thesaurus entries next to words like corruption, profiteering, racketeering and venality. In general, the whole messy predicament had turned Nigeria into an economic mess that to this day precariously teeters on the strength and weakness of oil prices. From this viewpoint, it is no wonder why the country has had the reputation it garners.

And yet, Nash would consistently reassure us that things were now different. That the country, and even detestable Lagos had its merits, its civility, and its comforts. That since 1999, the country had been mending albeit slowly with the first elected president residing over democratic government. He would tell us that the time was now ripe for science to return to his troubled homeland.

However, truth be told, I wasn't sure what or who to believe anymore. Frankly, my mind was now more preoccupied with evaluating the worth of my small science workshop. Nigeria, it seemed, had problems that were much bigger than biotechnology.

* * *

Murtala Mohammed International Airport in Lagos seemed to be as good a place as any to feel out the current truths on Nigeria's reputation. This place was to be our testing ground, a microcosm of sorts, and would provide us with a visceral sense of things to come. In this respect, the airport would be a focal point of a number of things. First up was the task of transporting scientific supplies from Canada into Nigeria. Part of my responsibilities entailed the delivery of some special chemicals that were not readily available in Nigeria. An innocent enough endeavour as these chemicals included relatively obtuse things like ultra pure varieties of water and salt; as well as some more sophisticated reagents like antibodies, DNA molecules, and even live but attenuated bacteria. A few of the tubes even contained samples of fake data which were a teaching lab's last resort should the experiments go completely awry. All of which were harmless and worthless of course, given that my work centered round educational goals rather than specific research goals. Unfortunately, all of the reagents needed to be kept cold in order to stay happy, which meant packing everything in copious amounts of dry-ice, a problematic venture given dry ice's hazardous reputation in the transport business. This nuance caused no end of frustration especially since the reason for it being labeled dangerous wasn't clear to me or, for that matter, to any of my chemistry colleagues.

Nevertheless, it was imperative that these supplies make it to Nigeria. For advice, I contacted Dr. Terry Pearson, a noted parasitologist at the University of Victoria, whom Nash had invited as the keynote speaker for our workshop. Terry, who specializes in Trepanoma research, the causative agent of sleeping sickness, has long had collaborations with various institutes in Kenya and was the person in my neck of the woods to consult on the ins and outs of transporting scientific goods to Africa. Although, we would later formally meet in Nigeria, he was a veritable wealth of information – a Yoda-like antidote to my ignorance.

"Don't even bother with a courier." he would tell me, in a gruff friendly manner that perfectly imbues someone equally at home in a state of the art biochemistry lab and in the safaris of southern Kenya. "They're too damn expensive, and in my opinion, you'll never see your package again." He added, "You can try the air cargo route, but I've never had much luck with that myself."

He then went on to explain that the only way to guarantee getting the materials over to Africa would be to simply bring the icebox to the airport and carry it on the airplane myself. "You've got to go to the airline agent and suck up big time, hope that they trust you, and let you bring it on. That's really the best way to do it. The only way to guarantee that you actually get it to your destination. If you want, you can bring all your fancy credentials – it certainly can't hurt. But at the end of day, it all depends on whose manning the desk, and whether they like the look of you or not."

It boggled my mind that my best chance in all of this chaos would be to bring the reagents onto the plane myself. Reagents that included among other things, an assortment of scary sounding chemicals, kilos of dry ice, and (get this) live bacterial samples. I shuddered at the thought of trying to explain to airline staff on the harmlessness of the bacteria in the security intense world of post September 11.

Consequently, I thought it pertinent to also give the airmail route a go using either a courier or cargo service. Terry was not kidding. Although the reagent package itself was no bigger than say a stack of 5 compact discs, the process of transport itself due to flight connections, and navigation through various custom ports, meant that the trip would take a minimum of 4 days. Which translated to an awful lot of dry ice to keep things frosty throughout the whole venture. This now heavy package would subsequently amount to a small fortune in shipping costs – to the tune of about US$450 via a company like Federal Express. or a more reasonable but still expensive US$200 using an air cargo service. Still, it was an alternative worth using, and approximately a week before the workshop, I arranged to send package number one into the unknown.

Package number two was safely bundled in one of those squishy ice bags, and packaged with enough dry ice to last for a 16 hour flight. It also came with a wad of paper, large enough to look like a novella, but which actually contained MSDS or materials and safety data sheets for each and every item in the package, as well as photocopies of every official looking letter of intent. The MSDS sheets were considered to be both advantageous and detrimental in our efforts to board the plane. They were the documents required to state the relative hazard level of each chemical, but in reality they often depicted the most harmless of things as being dangerous themselves – after all, even water has a lethal dose. As I approached the airline desk with Sam, it suddenly dawned on me that my chances of getting the package on board the plane would have increased significantly if I had my baby Hannah as a token gesture of innocence. However, in one of those surreal moments where one's expectations are flogged and hung, we experienced no problems, no lines of questioning, not even a customary look of concern. Neither Canada nor the United States seemed to care about my ice bag and my package number two was therefore on its way.

The actual arrival at the airport in Lagos was our second task, and was also an exercise in tension for both Sam and I. Tense, because according to all accounts, personal and otherwise, the airport was the place for peril and quite simply, it was the one part of the trip that we were most worried about. We were repeatedly warned not to converse with anyone that we didn't recognize, and to not take any mode of transportation without directions from someone we knew. Taxi drivers were notorious for partnering with any number of robbers and bandits, and it was reputed to be a calculated risk to interact with any stranger at the airport. Definitely not the most tourist friendly location I've been to. Basically, we were focused on thoughts of getting through without being detained or being robbed. In other words, we were under the uneasy impression that our lives depended on Nash showing up at the airport to be our personal escort.

Surprisingly, customs at Murtala Mohammed International Airport was brief, and my package number two caused no curiosity. In truth, the airport itself, seemed to be relatively quiet, no doubt a consequence of the armada of severe looking armed guards that manned all the key entrances. I could see through the main doors that the roadway outside was populated by a throng of locals, tidily kept behind a number of security fences. This was presumably to clear the main road of congestion, and possibly the riff raff we were told to avoid. For a fleeting moment, we sensed our folly at believing all of the tall tales, which at this point, seemed to be grossly exaggerated. Everything was going smoothly and everything seemed relatively civil.

However, as soon as we picked up our remaining luggage, we were accosted by two persons – a man and a woman – who aggressively asserted themselves as our escorts, and forcibly directed our cart towards the airport exit. Of course, this scared us out of our wits, as the very thing we had been warned about seemed to be happening before our very eyes. Even with our protestations, our new escorts were adamant on leading us into the outside throng.

Thankfully, before we ended up causing a minor scene, and likely one with the participation of armed guards, Nash appeared from the chaos to save us. It would be an understatement of monumental proportions to say that it was a relief to see him in person. We were suddenly empowered with the one criteria required for travel in Lagos – that is, a friendly face.
Boarding into a van, we soon found out that our two headstrong escorts were indeed official members of our host institution, the University of Lagos. In no small way, the whole incident clearly showed us the destructive power of heresy. Looking back, it's still hard for me to pinpoint whether our feelings of panic were attributed to information we had been given, or to the situation itself.

As we began to move, I immediately sensed a grimness in Nash's demeanor. He was not as animated as usual. When confronted he said, "The lab is below my expectations. The Local Organizing Committee has done a very terrible job of this." I, ever the optimist, bounced back, "Well, at least package number two made it through." I clutched the ice bag as if it was the most important thing in the world. "How did package number one fare?" I asked.

"It is here." he replied curtly, "but I cannot get it without paying a customs fee." When I explained to him that I explicitly made it clear in all the documentation that the materials were not of any worth, he just smiled at me and said, "Welcome to Nigeria."

* * *

Bribery, or dashing, as it is affectionately called in Nigeria, is a national pastime second only perhaps to soccer. Nash would explain to me that the airport authority controlling our cargo had asked for a flat fee of 30,000 Naira, which was roughly equivalent to US$300. A tidy sum of money that is no less than a small personal fortune in Nigerian standards. Of course, Nash, ever the pragmatic individual, flatly refused to pay. Over the next two weeks, we received daily updates on the current amount of dash needed to relinquish the package. On the last day, the bid fell to 5,000 Naira – to which Nash responded, "Send it back to Canada."

As predicted, our rendezvous at the airport provided a window to Nigerian life in the here and now. Sam and I both agreed that in hindsight, a lot of the apprehension we carried was a consequence of distorted information. What we saw instead, was a nation in transition, going from bad to not-so-bad (albeit slowly), and a culture that was as foreign as any other to the serious traveler.

It was when our van took us through the harrowed and crowded streets of downtown Lagos, that the full brunt of culture shock finally hit. Peering through the windows of our vehicle, we saw a completely different way of life – primitive, at times decrepit, and altogether frightening. The heat of the night seemed to mix well with the incessant dust and ramshackle nature of each and every building we saw. These sights made us instinctively think of home and so our first order of business was to look for an internet cafe. This was to qualm the fears of our loved ones with a friendly and reassuring message of "I'm in Lagos and go figure, I'm safe and sound." Then a welcome stop for food. I think Nash was trying his best to make us feel comfortable since he took us to a 'Mr. Biggs', which was more or less a clone of the MacDonald's variety, except with burly armed guards. As I ate, I found it decidedly ironic that the first meal I had on this exotic adventure was a hamburger and fries.

Nash then took us to our accommodations. We were staying at the University of Lagos Guest House which was more a hotel than a student residence and was considered luxurious at a cost of about US$40 a night. Although my own room was fairly large, this did little to obscure its general disheveled state. Despite this first impression, I counted my blessings since the room actually had reliable power in the form of a back-up generator, working air conditioning, a toilet (sort of), and a television with 5 channels (curiously, three of which were set to MTV). It even had malarial pest control in the form of staff spraying the room each night with a can of Raid, and it also came with a comforting set of door locks. In any event, it would more than suffice and I gladly dropped off my backpack, and then queried Nash for a good place to stash the contents of my ice bag.

Although we were very tired, Nash said that the best place would be in a scientific freezer, and that we should see the lab facilities immediately. The sense of urgency and the fact that he continued to look sullen was troubling to say the least. Here was the eternal optimist reduced to a stoic figure. I was thinking, "how bad could it be?" We had, after all, been corresponding with the Local Organizing Committee for several months, and according to all the pertinent check lists everything appeared to be sound. Everything should have been ready and waiting, and the only possible shortfalls we had to worry about, were the chemicals and reagents in my package number two.

* * *

The Local Organizing Committee or the L.O.C. for short was a group of scientists, and university representatives who were responsible for providing us the necessary facilities and equipment required for the laboratory workshop. However, it was clear upon first glance of our adopted lab space that the L.O.C. really had no apparent skill in organizing anything. To begin with, our allotted space was not even in the university campus but was instead situated several kilometres away at the Nigerian Institute of Medical Research (N.I.M.R.). Although this distance itself didn't sound so bad, the additional need to travel by car was very awkward given that the streets of Lagos consistently imbued a state of pandemonium. There was even a colloquial term for the act of driving – a very appropriate "go slow."

When we approached the iron gates of the N.I.M.R. compound, it was evident that the entire area was caught in the grips of a power outage. In the somber light of our vehicle's headlights, we drove further into the maze of buildings and stopped in front of a rather unpresuming looking wall. Then following flashlight beams, we were guided like airplanes up a flight of stairs and into a room that looked gutted, disorderly, and plain filthy. "This is our space," Nash said matter of factly.

In short, the space was horrendous, and dirty to the point that sterility would be pure fiction. In addition, the emergency generator backing our area was ranked somewhere between sub-par and non-existent, which meant that loss of power was a real consideration. However, this concern was soon countered with the realization that we were given next to no equipment to run our experiments anyway.

As we blew off the dust from the one solitary bench in the room, I was drawn to one of three conclusions. One, that Nigerian science had deteriorated to such an extreme that even a campus that boasted a student population of 30,000 students and connections with over 100 other Nigerian Universities, could not muster up enough basic equipment for 12 experimental stations. Two, perhaps Nash had been misleading us all along with respect to his country's resources. Or three, maybe all of those tall tales were not so tall after all.

The following morning was spent contemplating the Guest House breakfast menu, which basically consisted of eggs and toast, toast and eggs, eggs alone, or toast alone. The poor selection of items mirrored our facility and equipment status. Consequently, our second day in Nigeria would primarily concern itself with sorting these problems out and we would try to do this at the opening ceremony. Here, we would have the opportunity to meet face to face with some of the members of the L.O.C.. and voice our concerns. Unfortunately, it soon became very clear that the L.O.C. were largely indifferent. Instead, most of the members were more interested in garnering attention by using the ceremony as a chance to exhibit their skills in speech making. We did, however, have a few sympathetic ears – individuals who were quite apologetic about the whole situation and tried to explain that the lack of resources was a pressing reflection of the financial hardships within the research community. Not so much that the equipment didn't exist, but rather that the equipment was such a precious commodity that the distinction between the haves and the have nots was zealously guarded. In Nigeria, the notion of collaboration was swept under rugs, in favour of concealment and selfishness.

In retrospect, this reality was already evident, given our experience with transporting reagents. If securing a small shipment of chemicals had cost me several hundred dollars in transportation and "custom" fees, then what hope was there for the average Nigerian biologist. This was beginning to look less and less like a workshop on molecular techniques, and more and more like a Survivor episode for science geeks.

Fortunately, due to my background as a Boy Scout and an inbred predisposition to being prepared, I had taken every tall tale I heard in Canada to heart and had pretty much prepared for any and all contingencies. So after the opening ceremony, Sam and I immediately went to work assessing the general status of the workshop. To begin with, the sparse supply of equipment necessitated a change in how the students would work together. Nash had originally arranged for enough chemicals so that a class of 24 students could be split into 12 working pairs. However, the equipment that was given to us by the L.O.C. would limit us to only 3 working groups – each with 8 students, which was hardly optimal. Nash was especially angry with this nuance and even went so far as to accuse the L.O.C. of arranging this on purpose so that the excess reagents could be scavenged for personal use.

Even the equipment we did have was suspect at best. For example, the lab was supplied with one centrifuge that worked in a quirky fashion. This machine to the uninitiated is a device that contains a rotor capable of spinning around at high speeds. The net effect of this action is the creation of centrifugal force – the type of force that makes water stay within a swinging bucket. In essence, this machine speeds up the ability to separate constituents within a mixture according to density. For instance, a mixture of sand and water would result in a sand pellet at the bottom of a tube with the fluid on top. Anyhow, our centrifuge worked like a charm except that the lid wouldn't open unless you turned the machine upside down, which was a really aggravating feature since our separated samples would simply become mixed up again.

We were also in dire need of a Polymerase Chain Reaction machine or P.C.R. machine for short. This device essentially allows an experimenter to amplify DNA molecules such that you can have a billion fold increase of material to work with. In short, this makes it easier to observe and characterize DNA molecules, especially when the initial amount is very small. To the layman, P.C.R. was the token technique described in Jurassic Park. Although, not actually powerful enough to clone dinosaurs, its overall utility is such that this is the machine of choice for a number of key experiments such as DNA fingerprinting, DNA sequencing and any technique devised to look for specific genes. All in all, about a quarter of the procedures in the workshop syllabus would rely on this one piece of machinery. Frustratingly, the L.O.C. had actually provided us with a machine to use, but only after politely telling us that it was broken.

We had heard through the grapevine that the N.I.M.R. compound actually had a brand new P.C.R. machine, and soon found out that the device itself was only a few hundred feet away from our own disordered facilities. This adjacent building turned out to be a newly built structure with the explicit purpose of conducting research on the human immunodeficiency virus, the causative agent of AIDS. Funded by the Bill Gates Foundation, it was exactly the sort of space that our workshop should've been offered in that it was clean, it was fully equipped, it had reliable power, and from first sight, it even appeared to be underutilized. I immediately cursed my luck for it being a centre for HIV research since this would imply strict procedural rules that would make it next to impossible to borrow the entire facility itself – HIV is after all a hazardous organism. However, I saw no problem with the use of its PCR machine given that our samples were essentially benign and had no chance of jeopardizing any of the HIV work.

So all in all, pressing on with the workshop presented itself as a formidable task, and it was decided that clear objectives needed to be tackled. I would begin by assessing a general flowchart of how the lab would run, taking into account all of the possible shortfalls along the way. Sam would visit working internet to see if she could find information on ways we could cheat or MacGyver our way through certain procedures – essentially finding procedural tricks that would enable us to perform the experiments without the luxury of certain scientific supplies. And Nash would see about getting us permission and access to the Bill Gates' PCR apparatus. Taken together, it was clear that we would need some extra time before the lab was ready, so in order to stall, we announced that the first day of the workshop would be limited to a full day of lectures.

* * *

Meeting the students changed everything. In contrast to the wallowing apathy of the organizing committee, these young faces were a breath of fresh air. Although, it was a grueling first day of work where my time was spent rooted in front of a small blackboard, it was enlightening due to the enthusiasm of the young crowd. In light of all the hardships that this country faced, it was evident that the Nigerian youth took its education seriously. This was further espoused by the country's willingness to grant access to university level education based on free will alone. Despite this positive outlook, it was discouraging to ponder the future of bright individuals who appeared focused, willing to learn, and yet realistic in what outcome this workshop could bring. You also had students who naively embraced the hope that the workshop would literally change their views on science, and unquestionably lead to a better life. And of course, in a society such as this, we were not surprised by personalities who were desperate to leave the country, and simply viewed Sam and I with the ambitious and tenacious intent of a possible ticket out.

I paid careful attention to each of the student's attributes, because Nash had vested in me the responsibility of selecting the four most promising individuals. These four would then benefit from Nash's personal attention in mentoring and guidance. I took this decision very seriously because I really felt that being one of the chosen would result in a dramatic improvement in life itself.

All told, the practical sessions ran for two weeks and constituted a flash pan of memories. It was a session like no other, where each day brought a different obstacle to the mix, and like any challenge, there were notable ups and downs. The lab began in an interesting enough fashion with a brief visit from Dr. Emmanuel Denenu – a messenger from the Federal government and an unfortunate scapegoat for the frustration that must greet these students day in and day out. Although he came to discuss the future, opinions about the present took centre stage and the fire in the student's voices during this confrontation would continue to ring in my ears for the rest of the workshop.

On many occasions, my heart sank as I began to second guess the value of the workshop. Sometimes, it seemed all too silly to be coming to a place such as this, to teach something as absurd as molecular biology. These pangs of doubt were constantly highlighted by the looks of disappointment on the student's faces whenever it became all too clear that the chemicals and reagents that I lectured about, were simply not feasible in a country with corrupt shipping practices and freezers that were forcibly useless. How could one talk of delicate molecules, when we even had difficulty getting something as benign as chalk! (At one point, Nash would have to purchase some himself since the L.O.C. would refuse to help).

There was also a disturbing sense of innocence and a dangerous level of ignorance about the science in general. The students were often unrealistic in their assessment on what these genetic techniques could accomplish and how fast they could be performed. There were many instances where the idea of genetic modification was nonchalantly chosen as the be all and end all to experimental design. Many of the student presentations we heard would conclude with a simple "I would make this and that better by genetically modifying it." This eagerness to so readily adopt the use of molecular biology was troubling to say the least. In our own developed society, there is a general theme of caution applied to any scientific endeavour, and it was important that these students understood the possible ramifications and considered the ethical arguments behind the use of such technology.

We spent a fair amount of time deliberating these issues, and it was interesting to hear the musings and opinions of Nigerian students. Suffice to say, opinion was mixed in that whilst all agreed with the general intention of being ethically sound, several suggested that the barrage of ethical polemic was a western luxury – that only for countries as rich as Canada or the United States, which reflected a high degree of stability, was wasting valuable time and money arguing over the pros and cons of a technology acceptable. This was certainly food for thought.

As a whole, teaching conditions remained pretty primitive in that the equipment and facility dilemma never really sorted itself out. Still, we could boast a few tangible victories. For instance, we were able to secure access to the Bill Gates PCR machine, although this achievement took a little maneuvering and involved a fortuitous meeting with the director of the N.I.M.R. facility over coffee, crackers, and (presumably for public relation reasons) several photographs. We were extremely lucky with our power situation, in that it never directly affected our experimental procedures. Rather, the loss of power tended to conveniently coincide with the lecture components of the course – it was almost uncanny. Most astonishingly, we managed to tease success from about two thirds of our experiments, an almost unbelievable statistic that hopefully reflected our skills as instructors, our knowledge of the material, and of course, searing blind luck. In truth, we did have to resort to our "fake" data stash, but only once. It's also safe to assume that there are now 24 Nigerian students who are quietly pondering the identity of this MacGyver fellow that Sam and I kept talking about.

The intangible victories were a little harder to ascertain. Apart from the empirical act of teaching new material, I hope we were able to leave the students with a strong sense of what we thought was wrong with Nigerian science. That is the prevailing acceptance of the harsh but understandable 'every man for himself' attitude. For us, this philosophy was painfully evident from the very beginning in the effort required to get any help whatsoever. In our minds, this selfishness was shattering any hope of allowing Nigerian science to flourish. In simple terms, here we had a well educated society that was stifling under limited resources – it simply made more sense to us to foster collaboration, to rely and build on each other strengths, and to share valuable resources.

We also left behind a small legacy of embarrassment. Our negative experience with the Local Organizing Committee and with the University of Lagos, in general, bequeathed a lasting impression on the Federal Ministry of Science and Technology. Dr. Denenu was obviously disappointed with the behaviour of the L.O.C. In the interim, it seemed that more than a few eyebrows were raised during our workshop, and Nash consequently earned the right to formally interact with the government in all future programs. One can only hope that this will lead to better things.

Fittingly, the last day presented itself with two dominant and opposing memories. In an amusing turn of events, Nash and I were presented with a bill for various charges. It was stunning to us that the L.O.C. would have the audacity to ply us for even more money. The bill itself was a mockery with money attributed to items like extringencies, which, to this day, is one of the great mysteries of the workshop. To be perfectly honest, no-one was even sure what the word meant. Nash took care of the bill in his own special way, which is to say that he probably threw it away.

The other memory was that of the students huddled together in a welcoming circle, holding hands and saying a small prayer.

"Thank you lord for your generosity in life, that you hold together our faith in you, and that you provide what you can in our lives. Thank you for the opportunity in allowing us to attend this molecular biology workshop, that we might take what we have learnt to help our people and country in need. That we use it responsibly and with respect to our land. We ask that you guide us in shaping a better future. That you guide us in the spirit of collaboration. That you help us all of us stay true to this aim and to deny the spirit of selfishness. We asked this in the name of the Father, the Son and the Holy Spirit, Amen."

This is going to sound cliched but the workshop, and I suppose the whole experience itself, made me reflect on my own lot in life. I would be fooling everyone if I didn't comment that during each day of our laboratory and lecturing sessions, I felt a strong sense of relief and gratitude. Relief in that at the end of it all, I actually had the option of leaving it all behind. Gratitude because through some cosmic roll of the die, I was born in a country where food, shelter, and indeed a decent way of life were readily available. We really take too much for granted.

Today, Nigeria like most of my memories is slowly fading from my mind, in this case buried by the security of my own family and kept alive only in souvenir status by the occasion headlines that skim my consciousness. Where indeed does science fit in a country that is on the one hand dogged by controversy over Amina Lawal, a woman sentenced to lethal stoning by a Sharia judge; whilst on the other hand prepares for the Miss World pageant? Next time I talk to Nash, I'm going to have to ask him. Turns out, Nigeria didn't win the World Cup that year, but then again I suppose there's always next time.

WORDS I SEE WHEN I READ THE PHRASE "INTELLIGENT DESIGN" WHILE SQUINTING

Interior Design

Bullshit

www.ingramcontent.com/pod-product-compliance
Lightning Source LLC
Chambersburg PA
CBHW032014170526
45157CB00002B/686